Security and the Envi

In 1993 the first Clinton administration declared environmental security a national security issue, but by the end of the Bush administrations environmental security had vanished from the government's agenda. This book uses changing US environmental security policy to propose a revised securitisation theory, one that both allows insights into the intentions of key actors and enables moral evaluations in the environmental sector of security. *Security and the Environment* brings together the subject of environmental security and the Copenhagen School's securitisation theory. Drawing on original interviews with former key players in United States environmental security, Rita Floyd makes a significant and original contribution to environmental security studies and security studies more generally. This book will be of interest to international relations scholars and political practitioners concerned with security, as well as students of international environmental politics and US policy-making.

RITA FLOYD is British Academy Postdoctoral Fellow in the Department of Politics and International Studies at the University of Warwick, and Fellow of the Institute for Environmental Security.

Security and the Environment

Securitisation Theory and US
Environmental Security Policy

RITA FLOYD

CAMBRIDGE
UNIVERSITY PRESS

CAMBRIDGE
UNIVERSITY PRESS

University Printing House, Cambridge CB2 8BS, United Kingdom

Published in the United States of America by Cambridge University Press, New York

Cambridge University Press is part of the University of Cambridge.

It furthers the University's mission by disseminating knowledge in the pursuit of education, learning and research at the highest international levels of excellence.

www.cambridge.org
Information on this title: www.cambridge.org/9781107416642

First published 2010
First paperback edition 2014

A catalogue record for this publication is available from the British Library

Library of Congress Cataloguing in Publication data

Floyd, Rita, 1977–
Security and the environment : securitisation theory and US environmental security policy / Rita Floyd.
    p. cm.
  ISBN 978-0-521-19756-4 (Hardback)
    1. Environmental policy–United States.   2. National security–United States.
I. Title.
  GE180.F57 2010
  363.7'05610973–dc22

                                  2009051872

ISBN 978-0-521-19756-4 Hardback
ISBN 978-1-107-41664-2 Paperback

For (and with) Jonathan

Contents

Tables

Acknowledgements

This book was originally my PhD thesis and the list of people who have helped me with developing the ideas it contains is extensive. I thank them all, yet here I will only be able to list the most important ones.

Stuart Croft was my thesis supervisor and later my unofficial yet de facto mentor for the Economic and Social Research Council (ESRC) post-doctoral fellowship during which I completed this book. Stuart has been everything a supervisor should be and more, and I am deeply grateful for his help, advice and friendship over these past few years.

Roland Dannreuther was my Master's dissertation supervisor at Edinburgh University. It was he who suggested applying securitisation theory to the case of US environmental security. How this would pan out neither one of us could have imagined at the time, but I am grateful to both him and my second supervisor Chad Damro for discussing initial ideas on several occasions.

Ole Wæver and Ben Rosamond were my thesis examiners. My thanks to them for their extensive comments; they have proved most helpful. I should like to thank Ole also for making available many of his older unpublished papers without which this book wouldn't be quite what it is now.

I am grateful to John Haslam of Cambridge University Press for giving me this great opportunity. I am also grateful to two anonymous reviewers at Cambridge University Press for the extensive, critical and ultimately extremely helpful comments they provided.

I am indebted to the many people I have interviewed, without whose willingness and interest in the subject and in seeing this project completed none of this would have been possible. Geoff Dabelko, however, deserves special mention here. Not only was he extremely welcoming and helpful when I was in Washington DC for interviews back in 2005, but also he time and again shared his many invaluable insights into the many complex empirical issues at stake. Besides a few individuals that wanted to remain anonyms I interviewed and remain grateful to: Bruce Beard,

Deputy Director, Environmental Readiness Office of the Deputy
Under Secretary of Defense; Alex Beehler, Assistant Deputy Under
Secretary of Defense (Environment, Safety, and Occupational Health)
(2004–2008); Curtis Bowling, Director for Environmental Readiness
and Safety Office of the Secretary of Defense; Shah Choudhury, Acting
Director for Environmental Management Office of the Secretary of
Defense; Phillip Clapp, President National Environmental Trust;
Carole Dumaine, Deputy Director for Energy and Environmental
Security, Office of Intelligence and Counterintelligence, US Depart-
ment of Energy; Leon Fuerth, National Security Advisor to Vice
President Al Gore (1993–2001); Sherri W. Goodman, Under Secretary
of Defense for Environmental Security (1993–2000); Wendy Grieder,
NATO-CCMS staff officer in EPA's Office of International Affairs;
Abraham Haspel, Deputy Assistant Secretary for Energy, Environment,
and Economic Policy Analysis (1990–1999); Alan Hecht, Principal
Deputy and Deputy Assistant Administrator for International
Activities at the EPA (1989–2001); Laura Henze, National Sikes Act
Coordinator US Fish and Wildlife Service; Thomas Homer-Dixon,
Centre for International Governance Innovation Chair of Global
Systems at the Balsillie School of International Affairs; Carol Lancaster,
Deputy Administrator of the US Agency for International Development
(1993–1996); Claudia A. McMurray, Assistant Secretary, Bureau of
Oceans and International Environmental and Scientific Affairs
(2005–2008); Daniel Meyer, Director–Civilian Reprisal Investigations
US Department of Defense; Bill Nichols, Deputy Director Office of
the Deputy Under Secretary of Defense (Installations and Environ-
ment); William Nitze, Assistant Administrator for International
Activities at EPA (1994–2001); Jeff Ruch, Executive Director of
Public Employees for Environmental Responsibility (since 1997);
Dieter Rudolph, American Director of the Arctic Military Environ-
mental Cooperation (AMEC); Larry K. Smith, Counselor to the Secre-
tary of Defense Les Aspin and to the Deputy Secretary of Defense
William Perry; Richard Smith, former Deputy National Intelligence
Officer for Global and Multilateral Affairs (1994–1997); Robert
Smythe, Member Sierra Club International Program; Gary Vest,
former Principal Assistant Deputy Under Secretary of Defense for
Environmental Security (1993–2001); Larry Williams, Director of
Sierra Club's International Program; Timothy Wirth, Under Secretary
of State for Global Affairs (1993–1997); R. James Woolsey, Director

of CIA (1993–1995); Anthony Zinni, Former Commander-in-Chief of US Central Command (CENTCOM).

On the subject of interviews I would further like to thank Terry Terriff. It was he who, as a onetime co-supervisor, insisted on the utility of interviews and fieldwork at a time when I was really quite resistant. He was right for doing so. I thank Terry also for providing me with stacks of newspaper cuttings, articles and books on the subject of US environmental security.

Over the past few years I have presented several aspects of this book at conferences and workshops. A few of these events were particularly helpful. The first was a workshop on an earlier version of the completed draft manuscript kindly hosted by the Environmental Change and Security Program at the Woodrow Wilson Center, Washington DC in June 2008, generously sponsored by the ESRC. This workshop was attended by many of the former leaders of US environmental security including Sherri Goodman, Gary Vest, Abraham Haspel, Alan Hecht; as well as Dan Myers, Richard Smith, Jennifer Dabbs Sciubba, Arthur Bradshaw, Katrine Ebbesen Mehlsen, John Sewell, Gib Clarke, Karin Bencala and Geoff Dabelko. Most people had read all or most of the finished draft manuscript in preparation for the workshop, and I am deeply grateful for their many useful comments. Tina Davies from the University of Warwick, as well as Karin Bencala and Geoff Dabelko from the Woodrow Wilson Center, were instrumental in the organisation of the workshop.

Another memorable event was a lunchtime seminar hosted by the members of the Centre for Law, Philosophy and Public Affairs (CELPA) at the University of Warwick, who kindly discussed an earlier version of the final chapter of this book. I am grateful to all those who participated that day.

I have also benefited a great deal from presenting my work at the International Politics Research Seminar at the Department of Politics and International Relations at the University of Edinburgh, as well as to the Security Research Group at the University of Wales, Aberystwyth; my thanks to all the participants at these two events.

Aside from these events I would like to thank Jon Barnett, Simon Dalby, Hugh Dyer, Brian Smith, Richard Matthew and Matt McDonald for their comments on parts or all of the finished draft manuscript.

For financial support I am extremely grateful to the ESRC. They funded my PhD and later awarded me a postdoctoral fellowship.

I am especially grateful to Warwick University's research development officer, Katherine Branch, for her great help with my application to the ESRC postdoctoral fellowship scheme. Katherine also helped me with putting together a successful application to the British Academy's Postdoctoral Fellowship competition, and I would like to thank the British Academy for allowing me the time to put the final touches to this book.

Above all else, however, I am indebted to my husband Jonathan Floyd. The fact that I have found such a good home for what was once my doctoral thesis, indeed the fact that I completed this thesis, is down to him. He has helped me in more ways then I could say. He is on practically every page of this document; I hope the dedication reflects this.

Abbreviations

AMEC	Arctic Military Environmental Cooperation
ANWAP	Arctic Nuclear Waste Assessment Program
AOR	area of responsibility
APP	Asia–Pacific Partnership on Clean Development and Climate
BRAC	base realignment and closure
CAA	Clean Air Act
CCMS	Committee on the Challenges of Modern Society
CENTCOM	United States Central Command
CERCLA	Comprehensive Environmental Response, Compensation, and Liability Act
CIA	Central Intelligence Agency
CSLF	Carbon Sequestration Leadership Forum
DEIC	Defense Environmental International Cooperation
DEP	Defense Environmental Programs
DERP	Defense Environmental Restoration Program
DOD	Department of Defense
DOE	Department of Energy
DOS	Department of State
DRL	Bureau of Democracy, Human Rights and Labor
EISA	Energy Independence and Security Act
EPA	Environmental Protection Agency
EQ	environmental quality
ESA	Endangered Species Act
ESOH	Environment, Safety, and Occupational Health
ESTCP	Environmental Security Technology Certification Program
EWG	environmental working group
FFCA	Federal Facilities Compliance Act
FUDS	formerly used defense sites
FWS	Fish and Wildlife Service

GCC	Gore–Chernomyrdin Commission
GHG	greenhouse gas
INRMP	Integrated Natural Resources Management Plan
IPCC	Intergovernmental Panel on Climate Change
IR	International Relations
MBTA	Migratory Bird Treaty Act
MCA	Millennium Challenge Account
MMPA	Marine Mammal Protection Act
MOP	Memorandum of Partnership
MOU	Memorandum of Understanding
NATO	North Atlantic Treaty Organisation
NEPA	National Environmental Policy Act
NSC	National Security Council
NSS	National Security Strategy
ODUSD–ES	Office of the Deputy Under Secretary of Defense, Environmental Security
ODUSD–I&E	Office of the Deputy Under Secretary of Defense, Installations and Environment
OES	Bureau of Oceans and International Environmental and Scientific Affairs
OIA	Office of International Affairs
OMB	Office of Management and Budget
OSD	Office of the Secretary of Defense
OUSGA	Office of the Under Secretary of State for Global Affairs
PEER	Public Employees for Environmental Responsibility
PRM	Bureau of Populations, Refugees, and Migration
RCRA	Resource Conservation and Recovery Act
REPI	Readiness and Environmental Protection Initiative
SDREPA	Sustainable Defense Readiness and Environmental Protection Act
SERDP	Strategic Environmental Research and Development Program
USAID	United States Agency for International Development

Introduction

This book proposes a revision of the Copenhagen School's influential securitisation theory that both allows insights into the intentions of securitising actors, and enables the moral evaluation of securitisation and desecuritisation in the environmental sector of security. Securitisation theory holds that in international relations an issue becomes a matter of emergency politics/a security issue not because something constitutes an objective threat to the state or to some other entity, but rather because a powerful securitising actor argues that something constitutes an existential threat to some object that needs to be dealt with immediately if the object is to survive. The idea that by saying something, something is being done is, in language theory, known as a 'performative speech act'. In securitisation theory, however, the performative speech act part – the securitising move – only evolves into a complete securitisation at the point when a designated 'audience' accepts the speech act. Upon acceptance by the audience, the issue is said to have moved out of the sphere of normal politics and into the realm of emergency politics, where it can be dealt with swiftly and without the normal rules and regulations of policy making. As regards the concept of security, this means that it has no meaning outside of this logic; security is a 'self-referential' practice; the meaning of security is *what is done with it*.

The idea that security is a self-referential practice is not only the essence of securitisation theory; it is also the secret of the theory's popularity and its explanatory potency. More readily than rival security theories, it allows the security analyst to account for the essentially contested nature of security where one and the same concept may mean entirely different, and even opposing, things. Yet although this is a clear, strong point on the part of securitisation theory, the Copenhagen School's preoccupation with the idea that security is a self-referential practice brings with it two major shortcomings. First, when using securitisation theory an analyst is only able to study the '[self-referential] practice that *makes* something into security issues'.[1] Thus, the securitisation analyst studies who can securitise, on what issues, under what conditions, and with what effects,[2] whilst questions above and beyond the practice of security, such as those concerning the *intentions* of securitising actors (e.g. 'why do actors securitise?'), are ignored.[3] In this book I shall revise securitisation theory so that the analyst can account for the intentions of securitising actors. I argue that vital clues regarding the intentions of securitisers can be found in the identities of the *beneficiaries* of any given security policy.

Second, the idea that security is a self-referential practice allows no conceptual room for the theorising of what really *is* a security issue, nor for what *ought* to be securitised.[4] Under the Copenhagen School's

---

[1] Ole Wæver, *Concepts of Security* (Copenhagen: Institute of Political Science, University of Copenhagen, 1997), p. 48 (emphasis in the original).

[2] Wæver, *Concepts of Security*, pp. 14, 48; Buzan *et al.*, *Security*, p. 27.

[3] I use the word 'ignored' here deliberately as the Copenhagen School does not offer a theory for why actors securitise. Indeed, as I will show later on, the possibility of such analysis is actively rejected by Wæver. This said, for the sake of completeness it should be noted that at one point in their *Security: A New Framework for Analysis* (Boulder: Lynne Rienner, 1998, p. 27) Barry Buzan, Ole Wæver and Jaap de Wilde list 'why do actors securitise?' among the questions answerable by performing securitisation studies. They argue: 'Based on a clear idea of the nature of security, securitization studies aim to gain an increasingly precise understanding of who securitizes, on what issues (threats), for whom (referent objects), why, with what results, and, not least, under what conditions (i.e., what explains when securitization is successful?)' (*ibid.*, p. 32). That the 'why' is included here in the list of what can be done with securitisation theory does not, however, refer to the ability to theorise a securitising actor's intentions; instead it is a function of the fact that once one has analysed (a) who has done what with regard to security, and (b) what threats a given actor considers dangerous, one inevitably also learns about the reasons any given securitising actor holds for *why* security measures are considered necessary. In other words, the securitisation analyst's ability to answer the question 'why do actors securitise?' extends to nothing above and beyond simply replicating what the securitising actor said. The problems with this approach are examined in Chapter 2.

[4] In more detail, the Copenhagen School argues: 'Our securitization approach is radically constructivist regarding security, which ultimately is a specific form of social praxis. Security issues are made security issues by acts of securitization. [. . .] We do not try to peek behind this to decide whether it is *really* a threat (which would reduce the entire securitization approach to a theory of perceptions and misperceptions). Security *is* a quality actors inject into issues by securitizing them, which means to stage them on the political arena [. . .] and then to have them accepted by a sufficient audience to sanction extraordinary defensive moves' (Buzan *et al.*, *Security*, p. 204).

theoretical framework, the security analyst and the securitising actor are 'functionally distinct' entities with the security analyst in no position to assume the role of the securitising actor at any point of the analysis.[5] This, however, does not mean that the Copenhagen School feels 'obliged to agree' with any given securitisation.[6] On the contrary, the school holds strong views about the value of both securitisation and also of desecuritisation. They argue that, in all but a few circumstances, securitisations are morally wrong, whereas desecuritisations are morally right.[7] Notably, they arrive at these conclusions by way of what they take to be the effects or *consequences* of either action. In the case of securitisation, they take the consequences to be de-democratisation, depoliticisation, the security dilemma and conflict. In the case of desecuritisation, they expect politicisation, understood as a general opening up of debate. Although it is not the securitisation analyst's *aim* to bring about desecuritisation (unlike, for instance, the Critical security theorist who seeks to bring about emancipation or encourage self-emancipation), the securitisation analyst is potentially able, by providing insights into the effects of securitisation, to reduce both the scale and number of escalations and security dilemmas found in the world.[8]

The Copenhagen School anticipates that the securitisation analyst will arrive at the very same conclusions regarding the outcomes of securitisation and desecuritisation; this much is clear from the claim of

---

[5] *Ibid.* pp. 33–4.

[6] *Ibid.* p. 34.

[7] In the following I will use 'morally right' and 'morally permissible' interchangeably; and do the same with 'morally wrong' and 'morally prohibited'. Although the Copenhagen School does not use either one of these descriptions, their statements that 'securitization should be seen as a *negative*' whilst 'desecuritization is the *optimal long-range option*' show that this is what is meant (*ibid.* p. 29). In previous publications I have used their language of 'positive' and 'negative', as opposed to morally right and morally wrong, but I no longer see any reason to use this language and not the language of morality. I should further like to stress that although the Copenhagen School has this one-sided view of securitisation and desecuritisation they recognise that the mobilisation power unique to security can occasionally be put to good ends, and also that there is a certain attraction in this mobilisation power itself. They argue: 'In some cases securitization of issues is unavoidable, as when states are faced with the implacable or barbarian aggressor. Because of its prioritizing imperative, securitization also has tactical attractions – for example, as a way to obtain sufficient attention for environmental problems' (*ibid.* p. 29).

[8] *Ibid.* p. 206.

Ole Wæver (who is the originator of securitisation theory) that 'it is the duty of the securitisation analyst to point to the importance of desecuritisation'.[9] This is also what is meant when the Copenhagen School writes that: 'One of the purposes of [the securitisation] approach should be that it becomes possible to *evaluate* whether one finds it good or bad to securitize a certain issue'.[10] If, however, we take seriously the Copenhagen School's view that securitisation and desecuritisation are to be judged in view of their *outcomes*, then it is difficult to concur. Not only do securitisations not always lead to conflict and the security dilemma, but also, even if it is true that securitisation leads to the suspension of ordinary (democratic) politics, this is a morally wrong outcome only if we value democratic decision-making above everything else. If, for instance, we value the reduction of human wretchedness in the world above all else, then the suspension of ordinary politics is morally permissible, provided that human beings at large are the beneficiaries of security policies, and not power holders and elites.

In this book, by use of the example of the environmental sector of security, I intend to show that securitisations are not categorically morally wrong, but rather that, depending on the beneficiary of environmental security policies, securitisation can be morally permissible. Similarly, the finding that desecuritisation is categorically morally right holds only if desecuritisation always leads to politicisation – the Copenhagen School's anticipated outcome of desecuritisation. Again, using the example of the environmental sector of security, I shall show that desecuritisation does not always produce the same expected outcome, but rather can also lead to depoliticisation. Depending on the outcome, desecuritisation can then be either morally permissible or morally prohibited.

Considering that both the ability to theorise intentions and the ability to genuinely morally evaluate security policies are – or should be – essential parts of security analysis more generally, improvement on either front is clearly highly desirable. That being so, my aim in this book is to devise a stronger and even more compelling securitisation theory.

[9] This point was repeatedly made during the PhD training course 'Security Theory – Critical Innovations' in Copenhagen, Denmark (29 November–3 December 2004) conducted by Professor Wæver.
[10] Buzan *et al.*, *Security*, p. 34 (emphasis added).

Overview of chapters

This book consists of seven chapters. Chapter 1 offers a detailed analysis of securitisation theory. The chapter is structured into three parts. In the first part I examine in what way selected works by John L. Austin, Jacques Derrida, Carl Schmitt and Kenneth Waltz (the posited intellectual ancestors of the Copenhagen School) are relevant for securitisation theory. Above all, I argue that what is included in/ excluded from the theory (most notably for the purposes of this book, intentions) can be explained by going back to these thinkers. In the second part I analyse the meaning of 'post-structural realism' (Ole Wæver's self-described position) for securitisation theory. I argue that both securitisation theory's analytical strength and its normative weakness derive from this notion. Finally, in the third part I consider the coherence of securitisation theory over time and focus on the move away from state-centrism to a state-dominated field of analysis. I argue that this move is consistent with the logic of security according to securitisation theory. I further argue that as long as securitisations occur in practice, a basic core of securitisation theory remains useful into the future; which is why it is sensible to improve the theory in the various ways here suggested.

Chapter 2 develops my own revised securitisation theory. Contrary to the Copenhagen School, I argue that a securitisation exists, not when an audience accepts the existential threat justification, but instead when there is a change in relevant behaviour by the relevant agent, that is justified by this agent with reference to the declared threat. Securitisation then consists of two events: existential threat justification (the securitising move) and subsequent security practice. In a second step I lower the bar for the success of securitisation to a degree that securitisations are successful simply by virtue of existing, and not at the point when normally binding rules are broken and emergency measures are taken. The crucial idea informing this move is that even if a securitisation is not followed by the actions which constitute the Copenhagen School's criteria for success, the securitiser still has reasons for why they securitised; reasons that hold vital clues about the securitiser's intentions. In line with this, in a third step, I suggest that securitisations can take one of two forms, both of which allow insights into the intentions of securitising actors. The first form refers to those securitisations where what is done in the name of security (security practice) matches the rhetoric of the existential threat justification (securitising move). I argue that in such cases it is the intention of the securitising actor to secure the referent object of security they themselves identified as existentially threatened; I call this a 'referent object benefiting securitisation'. The second form refers to securitisations where there is a mismatch between the securitising move and the security practice. I propose that in such cases, it is not the referent object that benefits from the securitisation, but rather the securitising actor benefits, for example, by gaining a raison d'être, or by maintaining existing levels of funding. In such cases it is reasonable to suggest that an issue is or was securitised because of the benefits this has for the securitising actor. Correspondingly, I call this type of securitisation, 'agent benefiting securitisation'.

Existing securitisation theory is complete only with the concept of 'desecuritisation'. Desecuritisation is understood as the process whereby securitisation is reversed and formerly securitised issues are moved 'out of emergency mode and into the normal bargaining processes of the political sphere'.[11] In a fourth step I challenge this assumption as too simplistic, and I argue that this equation holds up only because the Copenhagen School works with such a wide definition of politicisation that desecuritisation almost invariably leads to politicisation. In its place, I set forth a narrower definition of politicisation as resting with official political authority only. On that basis it becomes possible to suggest that desecuritisation sometimes leads to depoliticisation.

In a fifth theoretical step I propose that my distinctions between (1) two types of securitisation in terms of who or what they benefit, and (2) two types of desecuritisation ('desecuritisation as politicisation' and 'desecuritisation as depoliticisation'), enable us to start thinking cogently about morally right and morally wrong securitisations and desecuritisations in the environmental sector of security.

Chapter 3 is the first of three empirical chapters that aim to test my revised securitisation theory on the example of United States environmental security from 1993 to 2009. It analyses the rise of environmental security in the US American context and locates the securitising move. Environmental security was first mentioned in the US National Security Strategy in 1991 and later under Clinton it became an intrinsic

---

[11] *Ibid.* p. 4.

part of US leaders/US government officials' vocabulary, and – as I will argue – the issue was subsequently securitised. Although the case of US environmental security under the Clinton administrations from 1993 to 2000 is regarded as the best-known example of environmental security in the relevant literature, a comprehensive analysis into the reasons of the Clinton administrations for securitising the environment remains outstanding.

Chapter 4 investigates what happened to environmental security besides rhetorical acknowledgement by relevant policy-makers. I shall show that the existential threat justification was not matched by security practice. Informed by my revised securitisation theory, I propose that the beneficiary of securitisation was not the stated referent object of security (in this case the American people), but rather it was the securitising actor itself that benefited from environmental security. This case study is thus an example of agent-benefiting securitisation.

In Chapter 5 I examine what happened to the various environmental security policies under the two George W. Bush administrations (hereafter Bush administrations) from 2001 to 2009. I argue that this case study is an example of 'desecuritisation as depoliticisation', because not only were the issues formerly part of environmental security no longer regarded as security issues by the Bush administrations, but they all but vanished from the administrations' political agenda.

Chapter 6 is the moral evaluation of securitisation and desecuritisation in the environmental sector of security. Extrapolating from the Copenhagen School I propose that securitisation has no intrinsic value; what matters are the consequences of securitisation alone. A focus on consequences corresponds to what in moral philosophy is known as a consequentialist ethic. Consequentialists hold that the right thing to do in any situation is to act with a view to maximising the best consequences.[12] What the consequentialist believes to be the best consequences depends, in turn, on the unit of value they endorse. On par with a majority of consequentialists I endorse human well-being as the highest value, and I argue that a functioning natural non-human environment is a necessary requirement for human

---

[12] Usually maximising, but there are exceptions, e.g. Michael Slote's *satisficing consequentialism* – the idea that an act is morally right if the consequences are good enough: Michael A. Slote, *Common-sense Morality and Consequentialism* (London: Routledge & Kegan Paul, 1985).

well-being. On this basis I am able to argue that only environmental security as human security is morally permissible, because only here are human beings the *beneficiaries* of the security policy. Other approaches to environmental security (environmental security as national security and ecological security) that promote different referent objects of environmental security altogether, are dismissed as morally untenable. As regards desecuritisation, the idea that the environment is instrumentally valuable to human ends means that only 'desecuritisation as politicisation' is morally right, as the safeguarding and protection of the global environment require political leadership at the highest level.

Chapter 7, which concludes the book, briefly considers the implications of this revised securitisation theory for security studies more generally and proposes a pathway for further research. I suggest that my analysis is relevant for the other sectors of security (identified by the Copenhagen School as military, societal, political and economic security) as well, because if there exist morally right and morally wrong securitisations and desecuritisations in the environmental sector of security, it is safe to assume that such distinctions exist in all sectors of security. Although my moral evaluation of securitisation and desecuritisation in the environmental sector does not suggest what these are for the remaining sectors of security, I argue that it offers some valuable insights into how to morally evaluate security policies in general.

1 | *The nature of securitisation theory*

Introduction

In Ole Wæver's many writings on securitisation theory there are three recurring as well as puzzling claims that have been left largely unexplained, even though each of them is vital for a comprehensive understanding of the theory. There is, first of all, the unconventional mix of theorists that are said to form the intellectual ancestors of the Copenhagen School. According to Wæver these include, besides John L. Austin and Jacques Derrida, Carl Schmitt and Kenneth Waltz.[1] Aside from the odd reference to one or more of these thinkers, it is, however, not clear what precisely is drawn from them and what in turn this means for the realm of securitisation theory. Consequently, there is some disagreement among those working on securitisation theory whether all of these thinkers are relevant. The conventional wisdom is that Waltz matters primarily for the Copenhagen School's regional security complex theory and the concept of sectors, and not so much for securitisation theory. This is contestable and in what follows I will draw out the important ways in which Waltz matters for securitisation theory too.

Second, Wæver (at least in his published work) has never satisfactorily explained what he means when he refers to his own alternative position as one of 'post-structural realism'. Considering the philosophical underpinnings of securitisation theory coupled with the lack of explanation one may wonder if this amounts to more than a mere amalgamation of labels. After all post-structural realism is a contentious position, considering that it seeks to combine two of the most opposed epistemological, ontological and methodological positions – poststructuralism and realism – into a unified position.

[1] Ole Wæver, 'Aberystwyth, Paris, Copenhagen: New Schools in Security Theory and the Origins between Core and Periphery', unpublished paper, presented at the International Studies Association's 45th Annual Convention in Montreal, Canada (2004), p. 13.

Third and finally, there is the issue of Wæver's changed opinion on
the role of the state in security analysis. Thus, in 1995, he forcefully
argued 'the concept of security refers to the state',[2] but only three
years later in 1998 he, as one of the Copenhagen School's joint authors
of *Security: A New Framework for Analysis*, makes the case for the
incorporation of other referent objects of security, including the individual. Does such a move constitute a fundamental contradiction in
the theory over time, or is this consistent with the theory?

This chapter clarifies each of these three unclear issue areas. In
order to do this I draw heavily on a number of unpublished papers
by Wæver, all of which, however, have been presented at international
conferences such as the International Studies Association's annual
convention. It is important to note that there is *nothing* in these
unpublished papers that contradicts the published work; use of
unpublished work merely allowed for a better and more comprehensive
understanding of securitisation theory and of the influences upon
Wæver's thinking in developing the theory. Overall, the benefit of
the analysis offered in this chapter is fourfold: first, it shows what
can and cannot be done with using securitisation theory; second, it
highlights the theory's strengths and weaknesses; third, it explains
what is included into the theory (excluded from it) and why; and
finally fourth, it aims to predict the longevity of securitisation theory.

The intellectual ancestors of securitisation theory

John L. Austin

'Security' is basically a speech act, or more precisely an 'illocutionary act'
[. . .]. Security is the sound coming forth when power-holders claim the need
to use their special right to block certain developments by reference to the
'security' of the state (or political order); a special right to use extraordinary
means going beyond their register in 'everyday politics'; a special right
grounded in the basic image of the modern state having the supply of
security and stability as its primary task.[3]

---

[2] Ole Wæver, 'Securitization and Desecuritization', in Ronnie D. Lipschutz, *On
Security* (New York: Columbia University Press, 1995), p. 49.

[3] This quote is taken from Wæver's essay 'Ideologies of Stabilization' published as
part of his 1997 PhD thesis *Concepts of Security* (Copenhagen: Institute of
Political Science, University of Copenhagen, 1997), p. 157. Please note, however,
that the quote itself is older than this essay. In footnote 1 it states: 'The first part

This statement is one of the earliest formulations of security as a speech act. To understand what exactly is meant here by a 'speech act', it will be necessary to explain the relevant work of John L. Austin, whose ideas on performative utterances (speech acts) form the basis of Wæver's thought. In Austin's posthumously published lecture notes, *How to Do Things with Words* (1962), the theory of speech acts is spelled out in meticulous detail. Inspired by Wittgenstein's language games, Austin argued that, until his time of writing, philosophy had only been concerned with 'statements' that can either be true or false and are necessarily descriptive. (Austin calls these constatives.) 'Statements' outside of the true/false dichotomy, namely those used to perform an action, have been ignored. Austin calls these performative utterances or performative speech acts. 'The name is derived, of course, from "perform", the usual verb with the noun "action"; it indicates that the issuing of the utterance is the performing of an action – it is not normally thought of as just saying something'.[4] In other words, performatives describe those speech acts where by saying something, something is being done. An example often used by Austin to show the working of the performative speech act is that of the marriage ceremony. He argues that the 'I do' uttered by bride and groom during the lawful marriage ceremony constitutes a performative speech act. In saying 'I do', something – the marriage – is being 'done'. '[T]he uttering of the sentence is, or is a part of, the doing of an action [. . .]'.[5] Another often used example by Austin is betting; alone by uttering the words 'I bet' something is being done – the bet has come into action.

Austin distinguishes between three different types of speech acts, namely, the locutionary act, the illocutionary act and the perlocutionary act. The most basic one is the locutionary act, whereby meaning is given to a certain utterance. An example of this would be: 'He said to me "Shoot her!", meaning by "shoot", shoot and referring by "her" to her'.[6] An illocutionary act goes beyond this and constitutes a meaningful

of the present study is an abridged version of a paper originally presented at the TAPRI workshop on Political Consequences of Nuclear Disarmament in Europe [in] 1988'.

[4] John L. Austin, *How to Do Things with Words* (New York: A Galaxy Book, Oxford University Press, 1965), pp. 6–7.

[5] *Ibid.* p. 5.

[6] *Ibid.* p. 101.

utterance coupled with a performative force. An example of this would be: 'He urged (or advised, ordered, &c.) me to shoot her'.[7] Third, a perlocutionary act is a meaningful utterance coupled with a certain force that brings about an unconventional effect. An example would be: 'He persuaded me to shoot her'.[8] Perhaps an easier way to illustrate these distinctions is made with this: 'We can [. . .] distinguish the locutionary act "he said that . . ." from the illocutionary act "he argued that . . ." and the perlocutionary act "he convinced me that . . ."'.[9] Securitisation theory makes use solely of the illocutionary act.

As already mentioned, performatives can neither be true nor false. They are, however, required to take place in appropriate circumstances and follow appropriate rules and regulations. Austin calls these 'felicity conditions'. 'Felicity' because, during the utterance, 'things [. . .] can be or go wrong' that make the speech act not 'false' but unhappy (infelicitous).[10] Austin lists six conditions that are required for performative speech acts to be felicitous. These are, briefly, first, the speech act must be in line with the 'accepted conventional procedure' referring to the utterance itself.[11] Second, 'the particular persons and circumstances in a given case must be appropriate for the invocation of the particular procedure invoked'.[12] Third, '[t]he procedure must be executed by all participants both correctly and [fourth] completely'.[13] Fifth, a person participating in a speech act must be sincere in her utterance. And sixth, the enunciator of the speech act must live in accordance with the utterance subsequently. Important to note about these felicity conditions is that not all of them are equally powerful. Instead, the first four are considerably more powerful then the two last felicity conditions. This is because the breach with any of the first four felicity conditions results in the 'misfiring' of the speech act.[14] Or, in other words, if these four felicity conditions are not observed, the speech act is *void*. In the words of Austin, 'a bigamist doesn't get married a second time, he only "goes through the form"

---

[7] *Ibid.* p. 101.
[8] *Ibid.* p. 101.
[9] *Ibid.* p. 102.
[10] *Ibid.* p. 14.
[11] *Ibid.* p. 14.
[12] *Ibid.* p. 15.
[13] *Ibid.* p. 15.
[14] *Ibid.* p. 16.

of a second marriage; I can't name the ship if I am not the person properly authorized to name it; and I can't quite bring off the baptism of penguins, those creatures being scarcely susceptible of that exploit'.[15]

Any breach with felicity conditions five and six, however, has less serious consequences, at least for the success of the speech act. This is because, regardless of the occurrence of insincerity whilst uttering the speech act and/or regardless of a breach with the promises contained in the speech act afterwards, the procedure can still take place and the speech act is *valid*. Such an abuse of the speech act, therefore, merely leads to the speech act's unhappiness, but not to its voidance. For example, if a person gets married without the needed thoughts and feelings required, and once married behaves in discordance with the marriage vows, then the speech act is unhappy, but the marriage is still valid and can only be fully dismantled by lawful divorce.

Directly derived from Austin's felicity conditions, Wæver identifies what he calls the 'facilitating conditions' of security as a speech act as follows:

1. The demand internal to the speech act of following the grammar of security and constructing a plot with existential threat, point of no return and a possible way out.
2. The social capital of the enunciator, the securitising actor, who has to be in a position of authority, although this should neither be defined as official authority nor taken to guarantee success with the speech act.
3. Conditions historically associated with a threat: it is the more likely that one can conjure a security threat if there are certain objects to refer to which are generally held to be threatening – be they tanks, hostile sentiments, or polluted waters. In themselves they never make for necessary securitisation, but they are definitely facilitating conditions.[16]

It is important to note that these 'facilitating conditions' do not pick up on Austin's distinction between those felicity conditions that render

---

[15] John L. Austin, 'Speech Acts and Convention: Performative and Constative', in Susana Nuccetelli and Gary Seay (eds.), *Philosophy of Language: The Central Topics* (Lanham MD: Rowman & Littlefield Publishers, 2008), p. 330.

[16] Ole Wæver, 'Securitisation: Taking Stock of a Research Programme in Security Studies', unpublished manuscript (2003), pp. 14–15.

the speech act void (conditions one, two, three and four), and those conditions that merely render it unhappy (conditions five and six). Neither do (what I choose to call) the 'sincerity' condition and the 'accordance' condition explicitly feature anywhere else in securitisation theory. Towards the end of the next chapter I will show, however, that their fulfilment is implicitly assumed by the Copenhagen School and that this has important repercussions for the question 'why do actors securitise?'

Jacques Derrida

The French literary theorist Jacques Derrida, (in)famous for 'deconstructing' works within western philosophy, analysed Austin's work on performatives in some detail in his 1982 essay, 'Signature event context'. In this essay Derrida argues that Austin's theory of performative speech acts, although innovative, falls short in the crucial aspect that it takes context as a fixed given. Such a fixed context and the presence of being is impossible, according to Derrida, for whom every utterance and every context is subject to a diffusion of meaning, a process in which the original context changes. In the terminology of Derrida, all contexts are subject to 'irreducible polysemia'.[17] Context can therefore never be fixed, but is always in flux. For the success or failure of performative speech acts, this then means that the performance of a speech act cannot be judged on experience, as, with context in flux, experience is irrelevant. Wæver is aware of this, and in his view there are the following consequences for securitisation theory:

[I]t is necessary always to keep open the possibility of failure of an act that previously succeeded and where the formal resources and position are in place (the breakdown of communist regimes in Eastern Europe) and conversely that new actors can perform a speech act they previously were not expected to (the environmental movement). [. . .] Therefore, the issue of 'who can do security?' and 'was this a case of securitisation?' can ultimately only be judged in hindsight. [. . .] It cannot be closed off by finite criteria for success.[18]

---

[17] Jacques Derrida, *Margins of Philosophy* (Chicago: University of Chicago Press, 1982), p. 322.
[18] Ole Wæver, 'Security Agendas Old and New – And How to Survive Them', unpublished paper, presented at the workshop on 'The Traditional and the

With regard to the previous discussion, it is important to note that Derrida's idea also shines through in the facilitating conditions. Condition number two, for example, explicitly states that the success of the speech act can never be taken for granted, whilst the careful circumscription of condition number three points to the Derridean/ poststructuralist influence, which makes it obligatory to avoid causality, essentialism and naturalism.

In addition, both the type of discourse analysis endorsed by securitisation theory and Wæver's belief that it is impossible, even undesirable, to theorise intentions of securitising actors results from his engagement with Derrida's work. For Derrida there is no meaning outside of the text itself. He argues:

[R]eading cannot legitimately transgress the text toward something other than it, toward a referent (a reality that is metaphysical, historical, psycho-biographical, etc.) or toward a signified outside of the text whose content could take place, could have taken place outside of language, that is to say, in the sense that we give here to that word, outside of writing in general.[19]

An easily comprehensible explanation of what this means exactly, including its consequences for analysis, has been offered by the historian of political thought Quentin Skinner. He argues:

[B]y far the most damaging campaign [to textual interpretation] was opened in the late 1960s and early 1970s by Jacques Derrida when he began to argue that the very idea of textual interpretation is a mistake, since there is no such reading to be gained. There are only misreadings, since it is an error to suppose that we can ever arrive unambiguously at anything recognisable as the meaning of texts.[20]

To exemplify the limits of textual interpretation, Skinner refers to Derrida's analysis of Nietzsche's sentence: 'I have forgotten my umbrella'. Here,

Derrida concedes that in this instance there is no difficulty about understanding the meaning of the sentence. [Yet] this still leaves us without any 'infallible way' of recovering what Nietzsche may have *intended* or meant.

New Security Agenda: Inferences for the Third World', Universidad Torcuato di Tella, Buenos Aires, 11–12 September 2000, p. 10.

[19] Jacques Derrida, *Of Grammatology* (Baltimore: Johns Hopkins University Press, 1998), p. 158.

[20] Quentin Skinner, *Visions of Politics I. Regarding Method* (Cambridge University Press, 2002), p. 91.

Was he merely informing someone that he had forgotten his umbrella? Or was he perhaps warning them, or reassuring them? Or was he instead explaining something, or apologising, or criticising himself, or simply lamenting a lapse of memory? Perhaps, as Derrida suggests, he meant nothing at all. Derrida's point is that we shall never know.[21]

For securitisation theory this is important in that 'it points to the centrality of studying *in* a text, how it produces its own meaning'.[22] That is to say, Wæver follows Derrida in his choice of discourse analysis. The latter is to be performed on publicly available texts only, which, in turn, limits the kinds of questions the securitisation analyst can pose and answer successfully. Given that he cannot hope to uncover what the securitising actor thinks, the question 'why do actors securitise?' is ignored and the securitisation analyst is to focus on who securitises, on what issues, under what circumstances and to what effect only. In Wæver's own words:

[S]ecurity thinking does not mean how the actors *think*, which would be rather difficult to uncover – and not all that interesting. What is up for discussion here is how and what they think *aloud*. That is, the thinking they contribute to the public debate/political process; 'public logic'. What we investigate is the *political* process – not the isolated, individual formation of ideas that are afterwards put into the political interplay.[23]

It should be noted here that, although Derrida has been singled out as the central influence behind these ideas, Wæver's hostility against theorising intentions can at least in part be attributed to the influence of his teacher and mentor, Ole Karup Pedersen. This is because Karup Pedersen in *Foreign Minister P. Munch's Conceptions of Denmark's Position in International Politics* (1970) highlights precisely the above logic, by denying 'the possibility of studying "what politicians really think"'.[24] For Wæver, who cites Karup Pedersen's 1970 book as one of the ten books that influenced his own development as a theorist the

[21] *Ibid.* p. 121 (emphasis added).
[22] Ole Wæver, 'The Ten Works', *Tidsskriftet Politik* 7 (2004), at www.tidsskriftetpolitik.dk/index.php?id=125 (11/2006).
[23] Wæver, *Concepts of Security*, pp. 116–17 (emphases in the original).
[24] Ole Wæver, 'Beyond the "Beyond" of Critical International Theory', unpublished paper, presented at the Joint Annual Convention of the British International Studies Association and the International Studies Association, London (1989), p. 73.

most, this was to become a significant revelation, later incorporated into his own theory.

[. . .] Karup Pedersen was particularly conscious to steer free of individual psychology. He took the full consequence and did not even try to ask what P. Munch 'really' thought 'deep inside', but what he found it opportune to state as his perception. [. . .] The dissertation did not ask 'behind' this to either what P. Munch 'really thought', nor to check it against reality [. . .] To use texts as a source to P. Munch's private perceptions would demand a non-existent psychological theory – there can be no valid analysis of individual psyche on the basis of the available sources. Therefore, Karup Pedersen elaborated a whole theoretical structure around the usefulness of understanding the presented conceptions as important in their own right.[25]

Nonetheless, it is the link to Derrida, which later gives the theoretical/ philosophical edge to this observation. We will encounter Derrida and of course the possibility of theorising intentions later on, for now I will continue with the work by twentieth-century legal theorist Carl Schmitt.

Carl Schmitt

In 'Words, Images, Enemies: Securitization and International Politics' (2003) Michael C. Williams identifies several elements of Schmitt's thinking that are important for securitisation theory. This article was one of the first to note the relationship between securitisation theory and Schmitt's work, a connection that has subsequently sparked much debate. Although Wæver came to accept the link between Schmitt and securitisation theory largely as a result of other theorists making this link, from his autobiographical 'The Ten Works' it is clear that he does accept it:

Some might expect to see Carl Schmitt on my list [. . .] It could almost be interesting whether I actually formulated the concept of securitisation with Schmitt in mind. Unfortunately, I do not remember. I was somewhat familiar with [Schmitt's] general argument, but as I recall, the original version of the speech act theory was formulated in 1988 without any direct inspiration. I only read Schmitt in detail later – and found him very convincing, noticing naturally the similarities, as well as the points where – hopefully – we part ways.[26]

[25] Wæver, 'The Ten Works'.
[26] *Ibid.*

Before examining these similarities in more detail, it will be necessary to briefly explain the essence and background of Schmitt's political theory.

Writing during the time of the Weimar Republic, the legal theorist Carl Schmitt argued contra the then orthodox legal positivists that the state could not be regulated by law alone, but rather by 'the political' in form of power and decision. The political is defined as 'the most intense and extreme antagonism, and every concrete antagonism becomes that much more political the closer it approaches the more extreme point, that of friend–enemy grouping'.[27] The reasons for this viewing of the political in terms of 'friend and enemy' can only be fully understood against the backdrop of the Weimar Republic, when government was made up of an array of political parties, including communist parties, which Schmitt regarded as extremist. In light of the representation of the 'extremist' parties in government, Schmitt saw the stability of parliamentary democracy as at risk because, without his concept of the political, extremists could, in a system based on law alone, make government unworkable, or worse, the constitution could simply be destroyed by a vote, should the 'wrong' party come into power. In order to obviate this, Schmitt advocated a strengthened position for the President Paul von Hindenburg. Here his concept of the political comes into play, in that he wanted von Hindenburg to name the extremists as the 'enemy of the state'. In Schmitt's view, von Hindenburg was in a position to do just that, because for him the sovereign decides upon the exception. He believed that the decision to name the extremists as the enemy of the state would catapult the chancellor on a long-term basis into a stronger position, with the rise of 'the political' forming a stronger basis for the continuation of parliamentary democracy.

Williams identifies two decisive connections between securitisation theory and Schmitt's thought. The first results from the existential threat requirement inherent in securitisation theory. That requirement means that: '[S]ecurity is not just any kind of speech-act, not just any form of social construction or accomplishment. It is a specific kind of act [because] it calls for extraordinary measures beyond routines and

[27] Carl Schmitt, *The Concept of the Political* (Chicago: University of Chicago Press, 1996), p. 29.

norms of everyday politics'.[28] As such, security 'mirrors the intense condition of existential division, of friendship and enmity, that constitutes Schmitt's concept of the political'.[29] In other words, just as the nature of 'the political' is determined by the division between friend and enemy, the nature of 'security' is determined by the division between normal politics where democratic rule is obeyed and extraordinary politics beyond rules and regulations.

The second decisive connection between securitisation theory and Schmitt's thought lies in what Williams calls Schmitt's 'decisionist theory of sovereignty'.[30] For Schmitt, the political – i.e. the distinction between friend and enemy – is strongest in the case of emergency, when the decision-making powers of the sovereign elevate the sovereign above the rules and regulations of the legal system. 'It is in the realm of emergency that the essence of sovereignty as decision is most clearly illustrated'.[31] According to Williams, this line of thought is clearly present in the process of securitisation, where a securitising actor is at its most efficient exactly because of operating 'legitimately' beyond otherwise binding rules and regulations.

Kenneth Waltz

The fourth and final thinker whose work constitutes an intellectual inspiration to securitisation theory is the International Relations scholar Kenneth Waltz. The first reason why Waltz matters for securitisation theory is of a meta-theoretical nature, that is, the first reason is not connected to what Waltz actually says about international politics per se, but rather what matters here is his view on the reach and, importantly, the limitations of theory.[32] For Waltz,

The construction of theory is a primary task. One must decide which things to concentrate on in order to have a good chance of devising some explanations of the international patterns and events that interest us. To believe

---

[28] Michael C. Williams, 'Words, Images, Enemies: Securitization and International Politics', *International Studies Quarterly* 47 (2003), p. 514.

[29] *Ibid*. p. 516.

[30] *Ibid*. p. 516.

[31] *Ibid*. p. 517.

[32] This, it is important to note, is *not* the same as saying that Wæver and Waltz have the same understanding of what constitutes a theory in International Relations, only that both thinkers agree that for a theory to have utility it needs to have boundaries.

that we can proceed otherwise is to take the profoundly unscientific view that everything that varies is a variable.[33]

For Waltz, therefore, certain things are by definition outside of any particular theory, for, if not, a theory becomes so watered down that it no longer has any explanatory power. In short, it ceases to be a theory. In Waltz's own version of realism outside of theory are, of course, the social and domestic make-up of units and how they interact. According to Wæver, Waltz has fundamentally changed the face of International Relations theory, importantly, however, not (alone) for what he actually says about the practice of international relations, but because of the standards he set for theoretical enquiry. Wæver writes:

> By emphasising the demands of social *science* and especially of *theory*, and taking a systemic-structural approach, realism was reformulated in a systematic and minimalist way, where previous broad speculations were replaced by a precise argument. The resulting theory was *theory* to a previously unknown degree in IR [. . .].[34]

Waltz's explanation of what a theory needs to entail and what is by definition outside of a particular theory is of vital importance for securitisation theory. Thus Wæver too subscribes to the idea that a theory cannot encompass everything and that a theory has to have boundaries. This fondness of 'Waltzian' boundaries of theory is evident in the following statement:

> To some security analysis functions as a *map of complexity*. One includes more sectors in order to be able to say that the military story is too simple, too narrow. And more actors serve to counter a state based account. [. . .] It becomes a kind of check list, a large matrix where one can put sectors along one side and units along the other, and then say: 'there are all these matrixes – see how the establishment only look at the small corner up there (at best 4 boxes), but there are 25'. In contrast, the focus on constellations and dynamics is aimed at reduction, at finding the turning points that might decide the way the future unfolds and thus function as a political analysis – even one that could be of help for political choices. The main difference is simply what kind of analysis one is interested in. Complexity versus constellations. [. . .] I am generally quite sceptical of [a focus on complexity] (as I am of much liberalist IR theory that only complicates matters in an attempt to

---

[33] Kenneth Waltz, *Theory of International Politics* (New York: Random House, 1979), p. 16.
[34] Wæver, 'The Ten Works' (emphases in the original).

give detailed 1:1 maps of the world instead of trying to simplify as realist theories at least are (more than) willing to.[35]

Context is one example of what for Wæver is outside of securitisation theory. Not only is context at odds with Derrida's role for the theory, for whom a fixed context and the presence of being is impossible, but also, according to Wæver, the inclusion of context (though perhaps drawing a clearer picture of the world than securitisation theory currently can) would change the theory beyond recognition, moving the focus away from the act that is securitisation, towards a causal theory of securitisation instead.[36]

In order to move on from this meta-theoretical level to the next two reasons why and how Waltz matters for securitisation theory it is first of all necessary to provide a brief sketch of Waltzian realism. In his magnum opus *Theory of International Politics* (1979) Waltz, in contrast to classical realists, provides a structural explanation of power politics, by arguing that the character of a state is largely irrelevant, as it is the structure of the system that constrains agents (states) and not the other way around.

One cannot infer the condition of international politics from the internal composition of states, nor can one arrive at an understanding of international politics by summing the foreign policies and the external behaviour of states. [T]he structure of a system acts as a constraining and disposing force, and because it does so systems theories explain and predict continuity within the system.[37]

Waltz identifies three layers that constitute the systemic (international) system structure: anarchy, individual unit capability and system capability (polarity). Polarity and unit capabilities are closely related in so far as the strongest units (the ones with the most capabilities) determine whether or not the system is unipolar, bipolar or multipolar. According to Waltz, states inevitably strive to balance each other's capabilities in an effort to minimise their own insecurity, whereby he conceives of security as coterminous with survival. It is feasible to suggest that Waltz's analysis has two direct repercussions for

---

[35] Wæver, *Concepts of Security*, pp. 366–7.
[36] Personal notes taken at Ole Wæver's presentation at the panel 'Critical Security Studies: Copenhagen and Beyond', International Studies Association 48th Annual Convention, Chicago, 2007.
[37] Waltz, *Theory of International Politics*, pp. 64, 69.

securitisation theory. First, for the Copenhagen School too, security is coterminous with survival.[38] Second, Wæver's and ultimately the Copenhagen School's restrictions on who is likely to succeed with a successful securitisation – 'the social conditions regarding the position of authority for the securitising actor'[39] – are modelled on a Waltzian, or more generally speaking realist, notion of the distribution of *capabilities* within the system. The more capabilities a securitising actor has, the more likely will this actor be to succeed in an attempted securitisation. In other words, who can or cannot securitise is already inscribed into the position of the actor within the social hierarchy of the system – in this book, government. This dynamic is captured well by Pierre Bourdieu who has criticised language theorists – including Austin – for not paying enough attention to the fact that the success of speech acts is always tied to the social position of the enunciator. For Bourdieu, 'power of words and power over words always presupposes other types of power'.[40] Not everyone will succeed in performing the speech act, but rather the 'capacity to make oneself heard, believed, obeyed and so on matters too. [According to Bourdieu this implies] that the efficacy of performance utterances is inseparable from the existence of an *institution* which defines the conditions (such as the place, the time and the agent) that must be fulfilled for the utterance to be effective'.[41] Michael C. Williams has recently usefully highlighted the importance of Bourdieu's concept of 'institution' for securitisation theory. He argues:

Power [. . .] emerges not only from the ability to speak the right language [. . .] but to do so in the context of having been accredited by institutions which have the power to confer that credibility. Thus trust and authority reside primarily not in the individual (at least not immediately, or per se), but more usually in the individual as mediated through their institutional accreditation.[42]

---

[38] Barry Buzan, Ole Wæver and Jaap de Wilde, *Security: A New Framework for Analysis* (Boulder: Lynne Rienner, 1998), p. 27.

[39] *Ibid.* p. 33.

[40] Bourdieu cited in Philippe Fritsch, 'Einführung', in Franz Schultheiss and Luis Pinto (eds.), *Pierre Bourdieu: Das politische Feld: Zur Kritik der politischen Vernunft* (Konstanz: UVK Verlagsgesellschaft, 2001), p. 10 (my translation).

[41] John B. Thompson, *Editor's Introduction*, in Pierre Bourdieu, *Language and Symbolic Power* (Cambridge: Polity Press, 1992), p. 8.

[42] Michael C. Williams, *Culture and Security: Symbolic Power and the Politics of International Security* (Abingdon: Routledge, 2007), p. 66.

To put the same point another way, different individuals enjoy different levels of influence within government and it is the level of influence which is crucial for the success of a securitisation. Thus whilst every actor can potentially be a securitising actor, success is dependent on the actor's capabilities. Put bluntly, a Vice President will find it easier to succeed as a securitising actor than a mid-level staffer.

Williams further argues that: 'Power, from this position, needs to be understood not in the materialist sense, but as symbolic power: the ability to use symbolic structures of representation and the occupation of social positions from which they can be effectively enacted, and social and material power thereby mobilized'.[43] Considering this, it is useful to suggest that, albeit inspired by Waltz, capabilities in securitisation theory do not exclusively refer to material capabilities, such as resource endowment, economic and military capability, as they do for Waltz; rather, capabilities are more usefully compared to Bourdieu's concept of *capital*, which contains besides economic (basically material) capital also cultural capital (knowledge, skills) and symbolic capital (authority). Just as Bourdieu's capital is different, depending on the field (political, cultural, literary etc.) in question, capabilities too are security sector specific. For example, whilst strong green credentials are a source of symbolic power in the environmental sector of security, they are of little use in the societal sector of security.

The meaning of post-structural realism

Post-structural realism I

Given these diverse philosophical underpinnings of securitisation theory, it is perhaps not surprising that Wæver refers to himself as a 'post-structural realist'. This label, however, refers to much more than a mere amalgamation of the philosophical underpinnings of securitisation theory; indeed, it seeks to offer a distinct approach to International Relations (IR) theory altogether.

The aim of 'post-structuralist realism' is to develop a theory that can be political. It does not want to find a new place to stand outside realism. Instead it tries to be at the limit of the tradition. [. . .] The method will be to work with

---

[43] *Ibid.* p. 65.

the realist concepts in order to mark them so that they are not able anymore to function in the harmonious self-assured standard-discourse of realism.[44]

According to Wæver, this aim is achieved by a double meaning of post-structural realism, as a 'poststructuralist reading of realism' on the one hand, and as a theory that is 'post' (i.e. after) Waltzian structural realism on the other. The first of these meanings is compatible with much of contemporary poststructuralist security studies, which takes the deconstruction of structural realism as its focal point, suggesting another way in which Derrida's work has been influential for securitisation theory. Deconstruction is a Derridean invention that refers to an 'event' within or occurring to a given text, whereby it is the objective of the person examining the text for the deconstruction to show

how texts based on binary oppositions themselves violate both the principle of exclusion and the principle of priority. Thus, a deconstructive reading of a text reveals points at which it introduces one of the opposing terms into the definition of the other or reverses the order of priority between the two terms.[45]

Structural, neo- or Waltzian realism is the particular target for post-structural IR theorists because for them realism (still perhaps the most prolific of all mainstream approaches) constitutes a selective reading of the history of political ideas, in that it allows for only one meaning of world politics, one congruent with *power politics*.[46] They argue, for example, that great thinkers on realism, including Thucydides, Machiavelli and even the realist icon Morgenthau, have been read and used to legitimise power politics, whereas passages that may well refute these thinkers' allegiance to the very concept have been ignored. It is the objective of poststructuralist IR theory to break this cycle of reinforced existence and to problematise the 'taken for granted' foundational elements of conventional IR theory: anarchy, sovereignty and

---

[44] Ole Wæver, 'Security, the Speech Act: Analysing the Politics of a Word', unpublished paper, presented at the Research Training Seminar, Sostrup Manor, 1989, revised, Jerusalem/Tel Aviv, 25–26 June 1989, p. 38.

[45] Gary Gutting, *French Philosophy in the Twentieth Century* (Cambridge University Press, 2001), p. 294.

[46] See, for example, Jim George, *Discourses of Global Politics: A Critical (Re) introduction to International Relations* (Boulder: Lynne Rienner, 1994); R. B. J. Walker, *Inside/Outside: International Relations as Political Theory* (Cambridge University Press, 1993).

the state. In utilising research methods, such as Derrida's deconstruction, poststructuralists have tried to dilute the boundaries between inside/outside and to give a voice to the 'other'. Wæver certainly is interested in, and has been influenced by poststructural readings of mainstream IR literature, in particular by the work of Richard Ashley.[47] However, when carefully examining Wæver's writings, it becomes clear that this influence has first and foremost amounted to the possibility of thinking critically about IR in general, rather than Wæver siding with post-structuralism completely. Consequently, Wæver differs from other poststructuralists in that he argues that the utility of the post-structuralist research project – in its predominant version – is ultimately unsatisfactory.[48] There are two reasons for this. First, he is critical of what (some) poststructuralists understand by 'thinking differently' about IR. He argues that the tendency in poststructuralist thought to ignore the traditional meaning of concepts leaves many new concepts underspecified, as it is only the difference to their traditional meaning that marks out new concepts. 'Truly critical analysis works with the logic of the traditional discourse – from the inside – and reinforces an already existing dimension of its internal logic until this becomes destabilising for the traditional security thinking. Then – because of the "respectful" treatment – the classical concept is displaced. We deal with the classical core – but in a new circumscription'.[49]

Second, because poststructural analysis is informed by the over-arching insistence on the ethical goal of 'opening up, making possible and freeing',[50] without anyone however, taking *responsibility* for what we are opening up to,[51] Wæver thinks that there is a naive belief underlying poststructuralist thought whereby the new is necessarily good. What if, so Wæver, we are opening up to something that is worse than what we had before? How can we possibly make 'opening up' the overarching aim, if we cannot know that opening up will make

---

[47] See, in particular, Ole Wæver, 'Tradition and Transgression in International Relations as post-Ashleyan Position', unpublished paper, presented at the British International Studies Association 15th Annual Conference at the University of Kent, 1989.

[48] Wæver, 'Tradition and Transgression', p. 37; Wæver, 'Securitization and Desecuritization', p. 86.

[49] Wæver, 'Security, the Speech Act', p. 37.

[50] Wæver, 'Securitization and Desecuritization', p. 86.

[51] Wæver, 'The Ten Works'; see also Wæver, 'Tradition and Transgression', pp. 35ff.

way for something that is good or better than what we had before?
Because we cannot, and because Wæver takes from Hannah Arendt
that we as human beings are inescapably part of politics, we have to
exercise a Nietzschean 'will to power' (which he understands as the
creation of values[52]), act *responsibly* and curtail our theorising in line
with learning from the past and with a view to the possible future.[53]
He further takes from Arendt that any position has to be judged by
the *effects* it gains in interaction with others, and therefore cannot be
'good' all on its own. Thus Arendt had stressed that there can be no
guarantee that our actions (or recommendations) will turn out to
have been correct, nor can it be guaranteed that our actions will be
understood, at least in the way we wanted them to be understood.[54]
Wæver translates this as follows:

> In the political, directions are not marked with road-signs indicating what is
> 'critical' or 'progressive'. In politics one cannot just be 'against': on the
> contrary one *does* something. Also as theory the critical practice moves *into*
> the political. It places its own respectability at risk in acts of which one
> cannot know whether posterity will re-tell them as confirmation or tran-
> scendence – as power politics or power criticism. Meaning comes after hand
> when the life story is told (Arendt 1958).[55]

And again a few years later: 'Acting politically can, consequently,
never be risk-free, and "progressiveness" is never guaranteed by one's
political or philosophical attitude. Theoretical practices, as well as any
political ones, have to risk their own respectability and leave traces,
letting posterity tell the story about the *meaning* of an act'.[56] And
more recently still, and here specifically with regard to the concept of
societal security:

> Whenever you let out a concept, you have to be prepared for it to take a life
> of its own. The concept of 'societal security' (i.e. identity security) easily
> lends itself to anti-immigrant, anti-EU and similar exploitation. One can

---

[52] In more detail, for Wæver the will to power is the 'willingness to stand up for
    projects and take responsibility as *creator*, not just as *conservator*'. Wæver,
    Concepts of Security, p. 176 (emphases in original); see also Wæver, 'Tradition
    and Transgression', p. 40, and Wæver, 'Beyond the "Beyond"', p. 46.
[53] Wæver, 'Tradition and Transgression', p. 40.
[54] Hannah Arendt, *The Human Condition*, second edition (Chicago: University of
    Chicago Press, 1998), p. 192.
[55] Wæver, 'Tradition and Transgression', p. 38 (emphases in the original).
[56] Wæver, 'Securitization and Desecuritization', p. 76 (emphasis in the original).

counter that this is a misunderstanding of the concept, but to some extent this misses the point. *One has to accept the power of words, including their power to liberate themselves from their creator.* Thus, the act of creating theory is as risky as any other political practice – you never know what harm you end up doing.[57]

This Arendtian logic of 'the unconscious of language, where the word travels independently of the logical structures one wanted to contain in it', coupled with Wæver's view of the limitations of poststructuralism, has had huge ramifications for his thinking; thus it is here that the idea originates that analysts must take responsibility for their own written and spoken words. Wæver does this by endorsing desecuritisation. By creating desecuritisation as a positive value, he hopes to take responsibility 'for what we are moving into' based on experiences, with the aim to curtail the research agenda he unleashed by virtue of having invented securitisation theory.[58]

To summarise, Wæver has been influenced by the idea or, better, the possibility of deconstruction only in so far as the possibility of the 'event' of deconstruction made him think beyond the horizon of mainstream IR theory, especially realism; nonetheless, with securitisation theory he offers no theory on how to do this; or rather on how to spot the event that is deconstruction.[59] With the originator of

---

[57] Wæver, 'Securitisation: Taking Stock', p. 29 (emphasis added).

[58] In Wæver's case, experiences refer very much to the role of détente in the ending of the Cold War, his research interest when he first developed securitisation theory and his alternative position, post-structural realism. He defines détente as 'negotiated desecuritisation; negotiated limitation of the use of the security speech act', which, so he argues, 'contributed to the modification of the Eastern societies and systems that made possible the radical changes of 1989' (Wæver, *Concepts of Security*, p. 227).

[59] It should be noted here that Wæver has used deconstruction, for example, in the essay 'Ideologies of Stabilization – Stabilization of Ideologies: Reading German Social Democrats' (article 5 in his 1997 PhD thesis *Concepts of Security*), which attempts to deconstruct the German SPD's security literature in the years before 1988. In this essay the aim of such deconstruction is 'not to create an explanatory, predictive model for SPD policy, but to learn something about the relationship between security, politics and text (language)' (p. 9). More important here for the purposes of this book is Wæver's observation towards the end of the essay which reads: 'What is to be learnt from this deconstruction? It is impossible to enclose a specific area of "security" and make unpolitical arguments from this basis. Security will always be drowned out by politics' (p. 175). A finding that leads him to conclude that security is always a political choice.

securitisation theory only partly subscribing to the poststructuralist research project, it is no wonder then that securitisation theory also is only partly poststructuralist.

Post-structural realism II

Moving on to the second meaning of post-structural realism, defined as 'a structural realism after Waltz', the definition of the timely element 'after' obviously is crucial. Thus what we are looking for here is a structural realism that builds on top of Waltz's initial framework that – it is imperative to remember – Wæver rates highly, but that supersedes the said framework in important ways. As we have seen, Wæver's thought was not only captivated by Waltz's ideas, but also by the ideas of Waltz's critics, most importantly those that showed a way of how to think critically about IR theory, the poststructuralist IR theorists.[60] Indeed, Wæver believes that 'one has to engage with traditions of thought, [as] theory works on theory'.[61] This suggests that the timely element of 'after' in this second meaning of post-structural realism refers to this process, the process whereby one theory emerges as a result of the critical engagement with another theory. In other words, not only does Wæver realise that post-structuralism emerged (not exclusively of course) out of a critical engagement with realism, his own approach to IR theory emerged from a critical engagement with *both* of these traditions. To understand why this is the case it is necessary to reconsider the following questions: What can be done/ not done with securitisation theory? What is the purpose of securitisation theory? And, finally, what is the securitisation analyst's method?

---

[60] To fully understand the importance of Waltz and his critics for Wæver, consider also the following quote taken from Wæver's list of the ten most influential books on his own development. The book discussed here is Robert O. Keohane (ed.), *Neorealism and its Critics* (New York: Columbia University Press, 1986): 'This item on the list is a cheat and in many respects unfair – first of all to Kenneth Waltz. Of course, his *Theory of International Politics* (1979) ought to have been on the list, but four of its central chapters are reprinted here, together with his important response to the critics. In addition, the book contains a number of key articles from debates on International Relations theory in the mid-1980s. This collection therefore conveys the importance of Waltz's text not only by way of the text itself, but simultaneously through the reactions that it engendered' (Wæver, 'The Ten Works').

[61] *Ibid.* See also, Wæver, 'Beyond the "Beyond"', p. 7.

Beginning with the second question first, the purpose of securitisation theory is to offer a theoretical tool of analysis with which the analyst can trace incidences of securitisation and desecuritisation as they occur in practice. Given this very clear-cut purpose the answer to the first question too is easily enough found; hence, in using securitisation theory the analyst must *not* focus on what security is, but rather on what security does, because what security *does* is tantamount to the meaning of security. Importantly by using securitisation theory the analyst cannot directly struggle against the securitisation/desecuritisation in question, even if she disagrees with the particular incident analysed. What is securitised/desecuritised is at the discretion of the securitising actor, not the securitisation analyst. To understand the complex relationship between actor and analyst it is worthwhile to cite the Copenhagen School here at length.

The designation of what constitutes a security issue comes from political actors, not analysts, but analysts interpret political actors' actions and sort out when these actions fulfil the security criteria. It is, further, the analyst who judges whether the actor is effective in mobilizing support around the security reference (i.e., the attempted securitizers are 'judged' first by other social actors and citizens, and the degree of their following is then interpreted and measured by us). Finally, to assess the significance of an instance of securitization, analysts study its effects on other units. The actor commands at only one very crucial step: the performance of a political act in a security mode.[62]

The securitisation analyst's method is one whereby she studies the discursive formation of the securitisation process in question. 'The way to study securitization is to study discourse and political constellations'.[63] Indeed, the way an analyst is supposed to study securitisations arises almost 'naturally' from the fact that security is conceptualised as a speech act, which means that the analysis does 'not entail conducting opinion polls and asking people what they think security means, but [rather the securitisation analyst studies] actual linguistic practices to see what regulates discourse'.[64] Stated as such, securitisation theory initially appears to concur with the poststructuralist research project. However, this impression changes (a) when we recall that securitisation

---

[62] Buzan *et al.*, *Security*, pp. 33–4.
[63] *Ibid.* p. 25.
[64] Wæver, 'Securitisation: Taking Stock', p. 9.

theory operates with a fixed and a rather realist meaning of security as survival,[65] and (b) when we account for the Copenhagen School's concept of 'inert constructivism', whereby it is argued that certain socially constituted entities 'can petrify and become relatively constant elements to be reckoned with'.[66] This is a view that enables the School to propose an approach that is at the same time 'constructivist all the way down' whilst they are able to study, for example, states or identities as almost objectivist entities.[67] It is these entities that are ultimately studied by using securitisation theory, with the aim to destabilise the traditional meaning of security, by showing that security does not refer to anything objective outside of itself; that it is a truly self-referential practice.[68] It is precisely this, the ability to examine traditional IR theory and security studies from within, with the purpose of the destabilisation of concepts endorsed by such theory, which defines the second meaning of post-structural realism and which sets it apart from the first meaning. Thus, although the necessity for destabilisation originates in the first meaning of post-structural realism, from that position, true destabilisation remains (for the most part) unsatisfactory, as without anchorage in the tradition wholesome destabilisation is impossible and new meanings of security are destined to remain simple mirror images of the tradition.[69] This is to say that those proponents of security that set forth alternative definitions of security overlook that even these alternative versions will lead to the same negative outcomes as traditional understandings of security, namely, security dilemmas and conflict, because security is 'by its nature a negative problem'.[70] In Wæver's words 'insecurity is the situation when there is a threat and no defence against it; security is

---

[65] A caveat here would be that although the meaning of security is fixed as 'survival in the face of existential threats' (Buzan *et al.*, *Security*, p. 27), the Copenhagen School accepts not only that the exact nature of these threats differs from sector to sector, but also that 'security issues are made security issues by acts of securitization. [Securitisation theory remains] radically constructivist regarding *security*, which ultimately is a specific form of social praxis' (Buzan *et al.*, *Security*, p. 204) (emphasis added).

[66] Buzan *et al.*, *Security*, p. 205.

[67] Barry Buzan and Ole Wæver, 'Slippery? Contradictory? Sociologically Untenable? The Copenhagen School Replies', *Review of International Studies* 23 (1997), p. 243.

[68] Wæver, 'Security, the Speech Act', p. 37.

[69] *Ibid.* p. 37.

[70] *Ibid.* p. 36.

a situation with a threat *and* a defence against it'.[71] In short, both security and insecurity are indefinitely tied to the threat–defence nexus; maximising security does not mean less insecurity.

Furthermore, precisely because the Copenhagen School holds that something becomes a security problem only when it is spoken of, or regarded as a security problem, they deny the objective existence of security threats. 'Things exist, but they do not come with labels on. It is a political choice whether something should be handled the security way or with "ordinary means" '.[72] This means that they believe that those security scholars that put forward alternative notions of security also do not point to any 'real' security threats; rather, with varying degree of success, they elevate problems to security problems. Informed by the idea that no objective security threats exist and in the anticipation that securitisation has negative consequences, Wæver and, by extension, the Copenhagen School have suggested that the only way to escape the threat–defence nexus is to achieve 'a-security'. A-security is defined as 'a situation that has been desecuritised or never securitised'; it 'is simply not phrased in these terms, [where] it is not a question of being secure or not and there is not a perception of existential threats being present'.[73] As a political strategy this takes the following form:

[W]e do not want to create a security theory that can only tell how everything could be different. [. . .] Transformation is one but not always the most reasonable strategy for improving security; in many cases, as analyst one can help more by grasping the patterns of action among units as they are and thereby help to avoid escalations, to steer vicious circles toward managed security complexes and eventually security communities.[74]

Analytical strengths and normative weakness

The idea that security is a self-referential practice is without doubt the Copenhagen School's single most important idea. This idea enables the securitisation analyst, better than other security analysts, to account for the essentially contested nature of security, where one and the same concept may mean entirely different, and even opposing things.

[71] Wæver, 'Securitisation: Taking Stock', p. 13 (emphasis in the original).
[72] *Ibid.* p. 20.
[73] *Ibid.* p. 13.
[74] Buzan *et al.*, *Security*, pp. 205–6.

That environmental security can mean as many different things as it did in the United States under the Clinton administrations, and many more, is therefore readily explained with the help of securitisation theory. Moreover, as regards the Copenhagen School more generally, this idea could be said to be the glue that holds together all of the work done by the School. Not only was it Barry Buzan, now one of the School's leading members, who highlighted the 'essentially contested' nature of the meaning of security in the sub-discipline of security studies,[75] but also regional security complexes and security sectors are the forums in which securitisation is possible and through which the complex dynamics of securitisation processes can be ordered. This much on the positive points; on a more negative note, the School's preoccupation with the idea that security operates in this way and no other, has led them to neglect what Thierry Balzacq has called 'external or brute threats'. These are threats that, in Balzacq's own words,

do not depend on language mediation to be what they are – hazards for human life. [In the Copenhagen School's] scheme, there is no security problem except through the language game. Therefore, *how* problems are 'out there' is exclusively contingent upon how we linguistically depict them. This is not always true. For one, language does not construct reality; at best, it shapes our perception of it. Moreover, it is not theoretically useful nor is it empirically credible to hold that what we say about a problem would *determine* its essence. For instance, what I say about a typhoon would not change its essence. The consequence of this position, which would require a deeper articulation, is that *some security problems are the attribute of the development itself.* In short, threats are not only institutional; some of them can actually wreck entire political communities regardless of the use of language.[76]

Although I am not in the business here of developing a theory of 'brute' threats, I think we can all appreciate that some things are threatening regardless of the use of language. Hitler's Nazi Germany,

---

[75] Barry Buzan, *People, States and Fear: National Security Problem in International Relations* (London: Harvester Wheatsheaf, 1983) and Barry Buzan, *People, States and Fear: An Agenda for International Security Studies in the Post-Cold War Era*, second edition (London: Harvester Wheatsheaf, 1991).

[76] Thierry Balzacq, 'The Three Faces of Securitization: Political Agency, Audience and Context', *European Journal of International Relations* 11 (2005), p. 181 (third emphasis added).

for example, was a real threat to its European neighbours during Prime Minister's Neville Chamberlain's time in office despite, tellingly, the latter's infamous failure to recognise this. Be that as it may, for our purposes here Balzacq's useful distinction is interesting primarily because it has negative repercussions for the viability of a political strategy of desecuritisation. Thus, in the case of brute threats, securitisation may well be a more viable political strategy than desecuritisation, considering that security's unique mobilisation power can deal with problems faster and more effectively than mere politicisation. This said, however, I do not wish to claim that the identification and subsequent targeting of brute threats by itself makes for a morally right securitisation. Instead, whether or not a securitisation is morally right depends on its outcome and, more specifically, on whether or not the securitisation improves human well-being.

Finally, provided that the arguments in this book are correct, a political strategy of desecuritisation is problematic as there simply is no guarantee that desecuritisation equates to politicisation and consequently no guarantee that desecuritisation is the morally permissible option. In other words, the potential existence of morally wrong desecuritisations shows that just pointing to the importance of desecuritisation – and irrespective of all good intentions on the part of the analyst – is not necessarily the most responsible way of acting. It is my claim that the security analyst acts infinitely more responsibly if he tries to distinguish between morally wrong and morally right desecuritisations and securitisations and thus shows potential securitising actors how to securitise and how not to securitise (and desecuritise). With regard to the environmental sector of security, this is of course part of what this book aims to do.

Coherence over time

From state-centric analysis to a state-dominated field

I now turn to the issue of whether securitisation theory has been coherent over the course of its short history and to the longevity of the theory. In the late 1980s, when securitisation theory was first developed, the state was almost universally regarded as a fixed element within security analysis. Considering that securitisation theory aimed to offer a tool for the analysis of incidents of securitisation and desecuritisation as they occur in practice, securitisation theory formed no exception to this trend. In 1989 Wæver wrote: 'Basically "security" is linked to the concept of "sovereignty". And to the idea founding the modern state – where it is for instance by Hobbes stressed how the first task is to secure *order*, domestic peace, stability of the political order'.[77] In the same article, Wæver, precisely to stress his point of the centrality of the state for security analysis, also critically engaged with Barry Buzan's 1983 book *People, States and Fear: The National Security Problem in International Relations*, which aimed at offering a comprehensive analysis of the national security problem and also introduced the individual level of analysis along with the international level of analysis to security studies. Wæver saw problems with both of these. For him neither entity was feasible as a provider of security because the ability to provide security lies with an actor's capabilities, something neither level possesses. Moreover, he considered neither well suited as referent objects of security. Individual security simply had no place in IR theory, whilst he thought it impossible to identify a referent object for international security, as 'there is no harmony between the different national goals [of different states]. [. . .] The meaning of "security" is basically "national security" and related to a specific problematique (the stability of the political order)'.[78]

In 1990, the year after this critical engagement with Buzan's book, *The European Security Order Recast: Scenarios for the Post-Cold War Era* was published, the first collaboration between, amongst others, Wæver and Buzan. And 1993 saw the publication of the second collaboration, *Identity, Migration and the New Security Agenda in Europe*. Although neither of these two books added anything of novelty to securitisation theory per se, the second book proved important for the development of securitisation theory's realm, because it was in this book, with the further development of Buzan's concept of societal security, that securitisation theory's hitherto rigid state-centrism first began to weaken. Thus, the 1993 book sought to offer a theoretical tool for analysis that could help theorise the changed European security landscape after the end of the Cold War. The book identified four areas of such change, all of which suggested a

---

[77] Wæver, 'Security, the Speech Act', p. 4 (emphasis in the original).
[78] *Ibid*. pp. 35–6.

new type of insecurity, one that no longer originated in inter-state conflicts, but rather one that resulted from the different needs and interests defining different European societies. The four areas of change were identified as follows:

1. The political stagnation and economic bankruptcy of the Soviet Union. 2. The revitalisation of Western European integration, initially under the banner of '1992', and latterly, and with much more trouble, under 'Maastricht'. 3. The widening acceptance that pluralism and markets were essential ingredients for any successful modern society. 4. The releasing and/or revival of nationalism and xenophobia.[79]

What was needed, therefore, was a tool for analysis that could help analyse these emerging societal insecurities. The key lay with the reformulation of Buzan's concept of 'societal security', so that that concept was no longer of interest only in so far as threats originating in the societal sector endangered the political stability of the state,[80] but it became about the capacity of a certain group to protect their identity.[81]

Whilst this was easy enough, things were more complicated when the School tried to conceptualise which groups have the necessary capabilities to securitise their identity. And also who, or what, therefore, provides societal security. Essentially what was needed was a societal entity that had the means to securitise and thus defend its existence against external threats. For the Copenhagen School the only societal entities based on a large enough 'we feeling' for securitisation to work were national identities and to a lesser extent religious movements. This hierarchy of identities is based on a strict division between what the authors call social groups and societal groups. Social groups – for example, environmental, sexual or ideological movements – are all valid identities, but by themselves are not strong enough to form a society, and therewith rival the state. In the hierarchy of identities, national identity was believed to hold the top spot over religion, because it can best unite multiple identities into a coherent whole, especially in times of crisis. Religious movements, on the other hand, were treated as subordinate to national identities,

[79] Ole Wæver, Barry Buzan, Morten Kelstrup and Pierre Lemaitre, *Identity, Migration and the New Security Agenda in Europe* (London: Pinter, 1993), p. 1.
[80] Buzan, *People, States and Fear: An Agenda*, pp. 122ff.
[81] Wæver *et al.*, *Identity, Migration and the New Security Agenda*, pp. 17ff.

because although they have 'greater flexibility of recruitment', religion – unlike nationalism – '[needs] to bridge the secular and spiritual worlds [with] religious identity [. . .]' not fully focused in the everyday world, and its spiritual elements can easily contradict, or not support, the political necessities of temporal existence'.[82] Here, it needs to be remembered that the theoretical framework of societal security in *Identity, Migration and the New Security Agenda in Europe* was developed with a view to the emerging European security landscape in the early 1990s and that it is in fact acknowledged that the basic 'societal units' need to be *rethought* for each case study and for each time period.[83] To put this differently, the above given view of religion is tied to the contextual background of the European Union in the early 1990s, and any deviation on this view regarding, for example, the importance of religion in later works does not constitute a contradiction, but instead a continuity in the Copenhagen School's work. Continuity, because all of the School's analysis is deeply tied to trends in the nature of security, with the respective analytical lenses and frameworks adjusted when changes occur.

Given that out of all possible societal groupings only national identities and religious movements support large enough numbers – and consequently have the necessary means for a securitisation to work – only these societal entities can provide societal security, whereby the latter is defined as: 'The ability of a society to persist in its essential character under changing conditions and possible or actual threats. More specifically, it is about the sustainability, within acceptable conditions for evolution, of traditional patterns of language, culture, association, and religious and national identity and custom'.[84]

Rejoining the argument above, it should be noted that societal security, despite being non-state-centric, did not make concessions to either individual or international security. 'Societal security offers an important extension to security theory. The elaboration of a concept of societal security has made clear that there is another social and collective focus in security analysis additional to the state, yet also standing in between the *unrealistic* extremes of individual and global referents of security'.[85] In other words, despite the acceptance of a

---

[82] *Ibid.* p. 22.
[83] *Ibid.* p. 190 (emphasis added).
[84] *Ibid.* p. 23.
[85] *Ibid.* p. 186 (emphasis added).

non-state-centric security concept Wæver did not contradict his earlier forceful argument of the impossibility of either international or individual security. In fact quite the opposite was the case and in 1995 much of the 1989 argument was reiterated in the now well-known chapter 'Securitization and Desecuritization' in Ronnie Lipschutz's *On Security*. Only three years later, in *Security: A New Framework for Analysis* (1998), however, the Copenhagen School's state-centrism was replaced by a conceptual openness to include other actors and referent objects beside the state. The reason for this broadening of analysis was as simple as it was powerful: new trends in the nature of security. In the decade after the end of the Cold War the role of the state, hitherto taken for granted, had become increasingly challenged by the emerging 'new world order' of the 1990s, whilst the concept of the state became simultaneously problematised in the academic world of IR theory. Given that the Copenhagen School is interested in practical security analysis their 'framework for analysis' needed to be able to adapt to these changes, particularly if it was to have longevity. That the Copenhagen School has been flexible enough to adapt to such changes from the outset becomes clear when considering the following quote taken from the School's 1993 book: 'there is nothing inevitable or permanent about defining security in state-centred terms; it has emerged historically and might change again'.[86] In the 1998 book, securitisation theory's realm was simply broadened to include other referent objects and providers of security than the state. Importantly, however, this book retains the argument that capabilities are instrumental for who can or cannot securitise successfully. This is exemplified very well in the treatment of the individual and the international level of analysis respectively, both of which, after having become issues in the policy-making world, now had to be incorporated into the Copenhagen School's framework, and could no longer be dismissed as non-existent, as they had been by Wæver in 1995. In other words, both the individual and the international were now seen as being possible levels of analysis. However, because of the problem of the lack of capabilities inherent to both of these levels, incidences of successful securitisation on either level were (at the time at least) few and far between. This finding led the Copenhagen School to suggest that although other referent objects are a

---

[86] *Ibid.* p. 20.

possibility, the state, the nation and civilisations were the most
prominent units of security analysis, precisely because these entities
had the capabilities needed for a successful securitisation. Even so, in
the hierarchy of referent objects the state still came top, because in
actual fact most securitisations, regardless of sector, still concerned the
state. The acknowledgement that 'security is an area of competing
actors' that is dominated by the state was called a 'state-dominated
field' of analysis,[87] a state of affairs – the Copenhagen School summar-
ises in yet a further book – that leaves securitisation theory to be 'not
dogmatically state-centric in its premises, but that is often somewhat
state-centric in its findings'.[88] In other words, there is nothing in the idea
of securitisation that ties it ultimately and definitely to the state.

Contradiction or continuity?

Whilst all of this may be self-evident to the securitisation enthusiast,
critics might not be so easily satisfied and rightly wonder whether this
change from a rigid state-centrism to a state-dominated field constitutes a serious contradiction in the Copenhagen School's writings.
Thus, although the change away from the state makes perfect sense
when considering that it is obviously what happened in practice, the
question that arises is: How does the securitisation theory, when
expanded in its scope, sit with the concept of security? Given, in
particular, Wæver's harsh words against widening the concept of
security along the referent object axis and his statement that 'the
concept of security refers to the state' (both in 1995), it must be asked,
first, how does he justify this revised version of securitisation theory?
And secondly, are the premises of securitisation theory still valid
following this change? Challenged by such questions, and by the
differences between his early and his later writings, Wæver has dedicated much effort to these problems in some of his work post the turn
of the millennium. The most significant contribution here is a conceptual history of security written in 2002. This history traces the concept
of security from (what he sees as) its beginnings in Roman times to
what we understand by security today. The analysis shows that the
concept of security shifted in several ways across centuries – for

---

[87] Buzan *et al.*, *Security*, p. 37.
[88] Barry Buzan and Ole Wæver, *Regions and Powers: The Structure of
International Security* (Cambridge University Press, 2003), p. 71.

example, from negative to positive, from subjective to objective etc. – making it impossible to determine the meaning of security as a constant – or as having a constant core.[89] The 1940s then bring order into the concept of security by joining the concept to the national interest. Ever since, so Wæver, what we understand by security is a product of the 1940s, with security modelled in accordance with *raison d'état* and necessity.[90] Importantly the widening of security has not changed this logic. Rather, despite the retreat of the state in issues of security and, more importantly, as a justification for urgent security measures (*raison d'état*), the *method* by which something is securitised, namely stating something in terms of absolute necessity, has prevailed, with this method now used by all sorts of securitising actors eager to ensure their own survival.

> The meaning the concept [security] now has is the speech act function to assign absolute priority to certain issues. They are *security* issues and therefore need to be dealt with before everything else, and if necessary without the usual constraints. There is no basis for this function of the concept in the previous history of security. The way the concept *works* is markedly shaped by the fact that it has taken over meaning from the other concept, raison d'état.[91]

In other words, the logic of the concept of security widened along the referent object axis remains the same as it was when only the state made up this axis. For securitisation theory this means that its internal logic, whereby a securitising actor claims the need for survival of a referent object, remains intact. It further means that a widened referent object axis can be accommodated *without* contradictions into the framework of securitisation theory.

Given all that has been said here, it should now be clear that there are two reoccurring issues that have shaped (and continue to shape) securitisation theory and its realm. These are, first, in order for a securitisation to work a securitising actor needs a certain amount of capability, as otherwise the securitisation will amount to nothing but a securitising move. Second, securitisation theory is aimed at studying

---

[89] Ole Wæver, 'Security: A Conceptual History for International Relations', unpublished paper, presented at the British International Studies Association's 29th Annual Conference in London (2002), p. 6.

[90] *Ibid.* p. 6.

[91] *Ibid.* p. 44.

securitisations and desecuritisations as they occur in practice. That the two are closely related, and that they are decisive, is highlighted well by the example of the recent, more thorough incorporation of religion into the Copenhagen School's framework of analysis. Whilst previously religion had been analysed solely in terms of how it relates to community (see above), by 2000 Wæver began to entertain the idea of religion as an independent sector of security, wherein religion was to be analysed for its own sake.[92] This move was heavily influenced by the twin realisations that (a) the securitisation of sacred objects by religious movements was not only possible, but for the most part easier than securitisation in other sectors, because the referent object for religion – *being* – appeals to the audience (here individuals of the same faith), for whom the survival of the referent object amounts to survival of the self,[93] and (b) that religion has become recognised as a security issue in practice.

In summary, although the move away from state-centrism to a state-dominated field of analysis appears to be a contradiction in the Copenhagen School's writings, it is in fact its opposite, a logical continuation of the School's thought. Informed by the joint forces of the significance of capabilities and that of the trends of security as they occur in practice, securitisation theory is in principle open to anything and everything that fulfil these two criteria. With security always in flux, changes such as these are part and parcel of the theory, and the ability to account for them is a major factor in the utility and the popularity of the theory. Given the ability of securitisation theory to adapt to changes in the nature of security, it is feasible to suggest that at least a basic core of securitisation theory will remain useful to security analysts into the future, which is but one reason why it makes sense to improve this theory further still.

Conclusion

It was the objective of this chapter to examine the nature of securitisation theory, including predicting its longevity. To this end it focused on three perennial puzzles accompanying the theory, namely: the

---

[92] Carsten Bagge Laustsen and Ole Wæver, 'In Defence of Religion: Sacred Referent Objects of Securitization', *Millennium: Journal of International Studies* 29 (2000), p. 709.
[93] *Ibid.* pp. 719–20, 739.

unconventional mix of intellectual ancestors of the Copenhagen School; the meaning of post-structural realism; and the changed role of the state in Wæver's writings over time. This chapter showed that select works by all four intellectual ancestors of the Copenhagen School, Austin, Derrida, Schmitt and Waltz, are invaluable for understanding securitisation theory as these are at the source of Wæver's deeply held assumptions on, amongst other things, intentionality, method and the limits of theory. It is possible to summarise the influences of the intellectual ancestors on Wæver's thinking in developing securitisation theory as follows. The idea that security operates like a performative speech act, whereby saying something is doing something, is taken from Austin's work on performative speech acts. From Austin, Wæver further takes the idea of facilitating conditions which he models on the former's felicity conditions. Whilst securitisation theory thus concurs with the idea that speech acts can never be true or false, for the wider purposes of this book it is vital to note that Austin's crucial distinction between speech acts that are void (misfire) and those that are merely 'unhappy' or infelicitous remains unobserved. From Derrida, Wæver takes the idea that context is never fixed but always in flux, which is to say that the conditions which would guarantee the success of the speech act cannot be theorised. From Derrida as well as from his former mentor Karup Pedersen, Wæver inherits the belief that we cannot know an actor's private thoughts. Consequently, the securitisation analyst is not supposed to 'try to get to the thoughts or motives of the actors, their hidden intentions or secret plans',[94] but rather studies publicly available texts only. From Williams we learn that the Copenhagen School's understanding of 'security' is tantamount to the Schmittian enemy and friend distinction of 'the political'. Wæver's understanding of 'the political', in turn, is taken from Arendt. The latter easily deserves a place as a fifth intellectual ancestor of securitisation theory – her awareness of 'the unconscious of language' encouraged Wæver to take responsibility for his own writings. From Waltz, Wæver takes his preference for parsimonious theory: the idea that security is about survival; plus, who can and cannot securitise is modelled on the Waltzian/realist idea of capabilities.

---

[94] Ole Wæver, 'Identity, Communities and Foreign Policy', in Lene Hansen and Ole Wæver (eds.), *European Integration and National Identity: The Challenge of the Nordic States* (London: Routledge, 2001), p. 26.

The analysis further shows that a mixture of select ideas by Derrida and Waltz are crucial to understanding Wæver's position of 'post-structural realism'. Post-structural realism harbours the purpose of securitisation analysis, which is to show that security is only ever a self-referential practice that refers to nothing real/objective outside of itself – that security is what is done with it. It is precisely this idea that leaves the securitisation analyst in an unprecedented position to account for the contested nature of security and study security policies as they occur in practice. The study of securitisations and desecuritisations as they occur in practice is believed to be of interest indefinitely and a basic core of securitisation theory, in turn, is expected to have longevity.

While clearly forming securitisation theory's strong point, post-structural realism, is simultaneously at the source of one of the Copenhagen School's biggest weaknesses – their one-sided view of the consequences of securitisation and desecuritisation. With this we have arrived at one of the most fundamental constraints of securitisation theory. The next chapter offers a revised securitisation theory that, amongst other things, paves the way to resolving this inadequacy.

2 | *A revised securitisation theory*

Introduction

The securitisation theorist's inability to say something meaningful about the moral value of different securitisations and desecuritisations is matched by his inability to theorise why actors securitise. Unlike the first shortcoming, however, the latter is an intentional choice on behalf of the Copenhagen School. Yet it is a choice that rests on an overlooking of the philosophical distinction between 'motives' and 'intentions'. To explain this, consider the following passage by Wæver: 'Discourse analysis works on public texts. It does not try to get to the thoughts or motives of the actors, their hidden intentions or secret plans. [. . .] What interests us is neither what individual decision makers really believe, nor what are shared beliefs among a population [. . .] but which codes are used when actors *relate* to each other'.[1] This passage quite clearly ignores that intentions are what an actor aims at or chooses to do, whereas motives are what determines an actor's aim or choice.[2] The distinction between the two concepts is vital, because whereas an analyst cannot get at what determines an actor's aim or choice (for instance, we cannot know for sure why Al Gore became interested in environmental issues as opposed to economic issues), we can know what an actor aims at in doing something (for instance, making the environment a security issue as opposed to not doing this). Let us consider another example, this time one adopted from the late Oxford philosopher Elizabeth Anscombe. If someone kills a person, they might have been driven by a particular emotion such as love or hate. These are the motives which led one person to take another's life. An outsider will have no means of knowing these, perhaps not even when the murderer

---

[1] Ole Wæver, 'Identity, Communities and Foreign Policy', in Lene Hansen and Ole Wæver (eds.), *European Integration and National Identity: The Challenge of the Nordic States* (London: Routledge, 2001), pp. 26–7.
[2] Elizabeth Anscombe, *Intention* (Oxford: Basil Blackwell, 1957), p. 18.

confesses to his crime. This does not matter, for motives in Anscombe's words 'are not causes at all'.[3] If I love or hate someone too much, it does not follow that I will murder the person in question. Murder only ensues when I wish to live in a world in which the loved/hated person does not exist, and when I am prepared to deal with the trials and consequences of murder.[4] In other words, intentions *are* causes. If they could not be known, intentional and unlawful killing (murder) could never even be detected.

This chapter will introduce a revised securitisation theory that, simultaneously, theorises intentions of securitising actors, and that paves the way for the moral evaluation of securitisation and desecuritisation in the environmental sector of security. Before any of this, however, let us begin with a look at some of the existing ethical criticism levelled at securitisation theory and the Copenhagen School.

Securitisation theory and ethical criticism

Much of the existing ethical criticism of securitisation theory focuses on the role of the analyst and his ethical responsibility in 'writing' or speaking about security. With regard to the Copenhagen School, there have been two interrelated issue areas which have been at the centre of this ethical criticism. The first of these is concerned with the absence of a normative or emancipatory concept within the analytical framework of the Copenhagen School, and the second is about the alleged disregard for the political consequences of the School 'writing' and 'speaking' security themselves, in their role as security analysts. I will begin with the latter.

Some of the ethical criticism of the Copenhagen School results from the concept of societal security, more specifically its referent object, identity, because 'the securitization of identity implies political risks and dangers'.[5] These political risks and dangers lie, of course, in the potential abuse of the speech act by groups with fascist, racist, xenophobic and similarly malign intentions, who by using securitisation, undermine the core values of the liberal democratic society. Critics of the Copenhagen School condemn not only this potential political

---

[3] *Ibid.* p. 19.
[4] *Ibid.* pp. 18–19.
[5] Johan Eriksson, 'Observers or Advocates? On the Political Role of Security Analysis', *Cooperation and Conflict* 34(2) (1999), p. 321.

outcome, but also the School's awareness of the possibility of such an abuse.[6] Based on the idea that the role of the analyst can never be neutral, Johan Eriksson argues that the Copenhagen School must take the political consequences of their teachings into consideration. He extrapolates his criticism from the Copenhagen School's claim that securitisation is a political choice, an argument he extends to the analysts themselves, claiming that Wæver *et al.*'s political agenda is visible in their aim to widen security studies. For Eriksson, securitisation theory is clearly at odds with the multisectoral approach to security. For Eriksson argues 'there is a contradiction between adopting a securitization perspective and not acknowledging one's own responsibility for widening the security agenda. The Copenhagen group seemingly want to retain as much as possible from their previous work, including the "sectoralization" of security, and to combine this with their more recently adopted securitization perspective. The two, however, are not compatible'.[7] The Copenhagen School is clearly political: 'If [. . .] seen from their own securitization perspective, they might be seen as acting more as politicians than as analysts, objectifying security, and spreading the negative connotations of threats and enemies to new issue areas'.[8]

Since Eriksson's critique was published as part of a symposium with, among others, Wæver, a direct response to the allegations raised is available. In his response, Wæver argues that Eriksson's critique takes the form of a logical contradiction in that 'we [the Copenhagen School] posit five sectors of security [. . .] as part of our analytical set up, i.e. before the finding of any study of actual securitizations'.[9] This claim, he continues, is untenable because:

[T]he presentation of a sector is not the same as a claim that there is such a thing as economic security or that it is widespread or legitimate. The set-up with five sectors is an analytical net to trawl through existing security

---

[6] For the Copenhagen School's self-awareness on this issue see, for example: Ole Wæver, 'Securitization and Desecuritization', in Ronnie D. Lipschutz, *On Security* (New York: Columbia University Press, 1995), pp. 65ff.; Ole Wæver, Barry Buzan, Morten Kelstrup and Pierre Lemaitre, *Identity, Migration and the New Security Agenda in Europe* (London: Pinter, 1993), p. 188.

[7] Eriksson, 'Observers or Advocates?', p. 315.

[8] *Ibid.* p. 316.

[9] Ole Wæver, 'Securitizing Sectors? Reply to Eriksson', *Cooperation and Conflict* 34(3) (1999), p. 335.

discourses to register what is going on. Whether we then find that there is lots of securitization in the environmental sector or not is a product not of the sectoral approach but of actors' practices. Given our preference (ceteris paribus) for de-securitization over securitization, *we would not bring up new sectors on our own initiative.* [. . .] The logical error is Eriksson's of mixing up a typology with claims about the empirical existence of the different types.[10]

That the selection of sectors has nothing to do with the Copenhagen School's personal preferences or own initiatives is ironically highlighted well by Eriksson's own example of the incorporation of the environment into the widened security agenda of the Copenhagen School. This is because both Wæver and Buzan are sceptical of environmental security as a solution to environmental problems. Their scepticism stems largely from a perceived conceptual mismatch between the differing degrees of intentional behaviour involved in the creation of threats. Violent threats have a high degree of intentional behaviour, and threats from environmental degradation have almost none.[11] Therefore, both Buzan and Wæver claim that the environment is not a security issue. Environmental threats occur irrespective of actors' wills; and for them it is precisely the question of actors' wills, or the idea of deliberate action, that establishes the structure for the field of security.[12] Following Eriksson's logic all the way through, surely they would then have excluded the environment from their sectoral framework. Yet, because securitisation of the environment is a real-life fact – as Eriksson himself points out – the incorporation into the framework was necessary and has been done; it simply has nothing to do with Buzan's and Wæver's personal choices, likes and dislikes. Or, as Max Weber said, 'only the adequacy of the data decides the question, which is wholly factual, and not a matter of principle'.[13]

Having dismissed Eriksson's critique as untenable, Wæver sets out to relaunch this very critique of incompatibility in a way that he

---

[10] *Ibid.* p. 335 (emphasis added).
[11] Daniel Deudney, 'The Case against Linking Environmental Degradation and National Security', *Millennium* 19 (1990), p. 464.
[12] Wæver, 'Securitization and Desecuritization', p. 63; Barry Buzan, 'Environment as a Security Issue', in Paul Painchaud (ed.), *Geopolitical Perspectives on Environmental Security*, Cahier du GERPE No. 92/05 (1992), Université Laval, Quebec, cited *ibid.*
[13] Russell Keat and John Urry, *Social Theory as Science*, second edition (London: Routledge & Kegan Paul, 1982), p. 199.

perceives as being 'much more painful to the Copenhagen School',[14] namely as a critique of the role of the political (social) scientist in co-constituting political reality. Here, the idea is that the securitisation analyst in writing (speaking) about a particular social reality is in part responsible for the co-constitution of this very reality, as by means of her own text this reality is (re)produced. For the Copenhagen School such a critique is defeatist, as their understanding of language makes any security utterance potentially securitising. Jef Huysmans has pointedly called this 'the normative dilemma of speaking and writing security'. He argues: 'Like a promise is an effect of language, that is, of successfully making the promise, a security problem results from successfully speaking or writing security. It is the utterance of "security" which politically introduces security questions in a publicly contested policy area. Thus, if successfully performed the speech act makes a security problem'.[15]

In other words, in writing or speaking security, the securitisation analyst herself executes a speech act; this speech act is successful if the problem raised becomes recognised as a security problem in the academic literature and/or in the wider policy-making discourse. The careful reader will have noticed that Huysmans' 'normative dilemma' is comparable to Arendt's logic of 'the unconscious of language' discussed in Chapter 1. It will come as no surprise then that the Copenhagen School's solution to both is, with the perennial endorsement of desecuritisation, the same.

This discussion leads on to the second ethical critique of securitisation theory, which is the often complained about absence of a normative, really emancipatory, objective within the theory. Indeed, given the preference for desecuritisation, many critics have taken to equating desecuritisation with emancipation.[16] The idea that emancipation is relevant for security has, above all, been developed by the so-called

---

[14] Wæver, 'Securitizing Sectors', p. 335.

[15] Jef Huysmans, 'Language and the Mobilization of Security Expectations: The Normative Dilemma of Speaking and Writing Security', unpublished paper, presented at the ECPR Joint Sessions, Mannheim, Germany, 26–31 March 1999, p. 8.

[16] See, for example: Hayward Alker, 'Emancipation in the Critical Security Studies Project', in Ken Booth (ed.), *Critical Security Studies and World Politics* (Boulder: Lynne Rienner, 2005), pp. 189–214; Richard Wyn Jones, 'On Emancipation: Necessity, Capacity, and Concrete Utopias', *ibid.*, pp. 215–36; Claudia Aradau, 'Security and the Democratic Scene: Desecuritization and Emancipation', *Journal of International Relations and Development* 7 (2004), pp. 388–413.

Welsh School of security studies who work within the tradition
of Critical theory. Critical theorists seek to emancipate or endorse
self-emancipation of humanity from what they see as the various false
and often dangerous consciousnesses of our orthodox concepts and
categories, whereby 'false consciousness' is the condition whereby
human agents 'falsely objectify their own activity'.[17] In line with these
ideas, the Welsh School views orthodox readings of security, whereby
security is coterminous with state and military security, as a manifesta-
tion of false consciousness in security studies and in the practice of
security. In their view, an understanding of security that is tied to
notions of 'power' and 'order' fails to make anyone truly secure, as it
constantly reinforces the security dilemma, whereby one actor's security is another actor's insecurity. They believe that true security can only
be achieved if ideas of security shift away from a focus of state and
military security to the security of individuals. For them true security
refers to the emancipation of the poor and the disadvantaged from the
way they live today. The Welsh School thus puts forward alternative
readings of political and social reality, whereby existing dichotomies of
security and insecurity are broken and where true security as emancipation can be achieved.[18] Richard Wyn Jones, who is a prominent
proponent of this school, makes the link between emancipation and
the Copenhagen School, suggesting that it has to do with a Habermasian reading of desecuritisation.[19] In a 2004 conference paper, Thomas
Diez and Atsuko Higashino have argued along these exact lines and
enlighten us how Habermas, desecuritisation and emancipation can be
combined. They argue that whilst securitisation closes down political
debate, desecuritisation opens up political debate, thereby moving
closer to a Habermasian 'ideal speech situation' – 'the situation in
which argumentative behaviour prevails over strategic behaviour'.[20]

[17] Raymond Geuss, *Idea of a Critical Theory* (Cambridge University Press, 1981),
p. 14.
[18] Ken Booth, 'Security and Emancipation', *Review of International Studies*
17 (1991), pp. 313–26; Ken Booth, 'Security in Anarchy: Utopian Realism in
Theory and Practice', *International Affairs* 67 (1991), pp. 527–45; Ken Booth,
Theory of World Security (Cambridge University Press, 2007).
[19] Richard Wyn Jones, *Security, Strategy, and Critical Theory* (Boulder:
Lynne Rienner, 1999), p. 123.
[20] Thomas Diez and Atsuko Higashino '(De)Securitisation, Politicisation and
European Union Enlargement', unpublished paper, presented at the British
International Studies Association 29th Annual Conference, University of
Warwick, 2004, p. 3.

In such a case, desecuritisation itself – as the absence of a world framed in terms of security – is the emancipatory ideal.

Another interesting possibility how securitisation theory can be thought of in accordance with a normative objective comes from Michael C. Williams as part of his above-mentioned study on Schmitt and securitisation theory. This normative objective is a result of securitisation being first and foremost a speech act, whose success is dependent on 'discursive legitimation'. With this he highlights the role of the audience for the process of securitisation. In citing Buzan *et al.*, Williams underlines his own argument for the importance of the audience: '[S]uccessful securitization is not decided by the securitizer but by the audience of the security speech-act: Does the audience accept that something is an existential threat to a shared value? Thus, security (as with all politics) ultimately rests neither with the objects nor with the subjects but *among* the subjects'.[21]

With security decided among the subjects, Williams sees a possibility for 'argument', which leads him to suggest that securitisation entails a commitment to Habermasian communicative action and discourse ethics. For Williams, by means of these elements, the Copenhagen School avoids Schmittian *realpolitik* logic and instead offers a 'political ethic', whereby securitisation is susceptible to the power of argument in the public and political realm.[22] Although an interesting idea, it could be argued that Williams (along with many others) is overly optimistic of the role of the audience in the act of securitisation. After all, the whole idea that securitisation is an 'intersubjective process' between the securitising actor and the audience is at odds with Waltzian and Austinian ideas about power, capability and performativity that inform the theory. As Thierry Balzacq has very convincingly shown, the idea that the process of securitisation is an illocutionary speech act – a one-directional performative whereby by saying something, something is being done – is at odds with the idea that security is an intersubjective process, conceived in the interaction between securitising actor and audience.[23]

---

[21] Barry Buzan, Ole Wæver and Jaap de Wilde, *Security: A New Framework for Analysis* (Boulder: Lynne Rienner, 1998), cited in Michael C. Williams, 'Words, Images, Enemies: Securitization and International Politics', *International Studies Quarterly* 47 (2003), p. 523.

[22] *Ibid.* p. 524.

[23] Thierry Balzacq, 'The Three Faces of Securitization: Political Agency, Audience and Context', *European Journal of International Relations* 11 (2005), pp. 175ff.

Another problem with the audience in the original securitisation framework is that it is difficult to ascertain who the audience is supposed to be. In Wæver's formulation the audience is not made up of the entire population, but rather 'it actually varies according to the political system and the nature of the issue'.[24] It appears that the most likely groups to be involved, or rather to be 'convinced', in cases of national security, are the political elite and also military officials. In the case of the US one possible candidate for the audience, then, is the US Congress, complete with its 'power of the purse'.[25] Such a claim, however, is difficult to sustain. After all, the Congress is part of the US foreign policy executive and as such surely part of the securitising actor. Furthermore, although the Congress can in principle assert its power and deny funding, 'the nature of international security issues and the increasing complexity of international politics make it difficult for Congress to lead the nation or, for that matter, to check the president on policy initiatives, especially during national crisis'.[26]

The most fundamental problem with the formulation of the audience in existing securitisation theory, however, is that it is not an analytical concept at all, but rather a normative concept in analytical disguise. This is because through the idea of 'intersubjectivity' Wæver can show, and indeed shows, a normative commitment. Inspired by Hannah Arendt's writings Wæver believes that politics should be done consensually and through dialogue and deliberation, as opposed to politics being a top-down process. Consequently, for him security policy too is viewed as an intersubjective process and is not decided by an individual actor or body.[27] In more detail: '[I]n order to avoid simply moving from objective to subjective – it should be stressed that since securitisation is never (in contrast to Schmitt) decided by one sovereign subject but in a constellation of decisions it is ultimately inter-subjective (and truly political in an Arendtian sense)'.[28] In other

---

[24] Ole Wæver, 'Securitisation: Taking Stock of a Research Programme in Security Studies', unpublished manuscript (2003), p. 12.

[25] Eugene R. Wittkopf, Charles W. Kegley and James M. Scott, *American Foreign Policy*, sixth edition (London: Thomson Wadsworth, 2003), p. 420.

[26] Sam C. Sarkesian, John Williams, John Allan and Stephen Cimbala, *US National Security – Policy-makers, Processes and Politics*, third edition (Boulder: Lynne Rienner, 2002), p. 193.

[27] Ole Wæver, *Concepts of Security* (Copenhagen: Institute of Political Science, University of Copenhagen, 1997), p. 3.

[28] Wæver, 'Securitisation: Taking Stock', p. 14.

words, the concept of the audience arises from Wæver's view of what politics *ought* to be, therefore not necessarily from how it actually *is*. Note here that the absence of the concept of 'the audience' from Wæver's early work is not a problem for my explanation; rather, it is a result of the fact that his thinking on the topic at that time was still very much 'work in progress'.[29]

Although the audience is not an analytical concept, in questions of national security there is a way in which what might be called 'the audience' matters for securitisation and for who can securitise. This is the case when 'the audience' confirms or rejects specific persons as political agents in a governmental structure that enables them to speak security more successfully than other actors. In liberal democracies this is usually done through elections. In this respect my idea of the securitising actor/audience relationship is very close to the politician/voter relationship in Bourdieu's political field. For Bourdieu the political field is distinct from any other fields (cultural, literary, mathematic etc.), precisely because politicians (as the main actors in the political field) in liberal democracies are not completely autonomous, but must, at regular intervals, present themselves to the populace by way of elections. 'One of the biggest differences between, for example, the literary field or the mathematical field and the political field is that politicians are subject to the discretion of the electorate. [. . .] They must, at regular intervals, face the voters who issue their mandate. As such, at least part of their action will always be geared to the public. Politicians cannot dream of complete seclusion'.[30] Of course, that same populace is free to demonstrate and the like against governmental securitisations as they take place, but ultimately, its most powerful tool is that of election and re-election.

Equating the audience with the electorate has several advantages. First, it accommodates Williams' idea of a political ethic whereby securitisation is susceptible to the power of argument in the public and political realm, precisely because who can securitise is determined

---

[29] For the role of the audience in Wæver's early writings and later work by the Copenhagen School see Holger Stritzel, 'Towards a Theory of Securitization: Copenhagen and Beyond', *European Journal of International Relations* 13 (2007), p. 363.

[30] 'Pierre Bourdieu im Gespräch mit Philippe Fritsch', in Pierre Bourdieu, *Das politische Feld: Zur Kritik der politischen Vernunft* (Konstanz: UVK Verlagsgesellschaft mbH, 2001), p. 34 (my translation).

by the electorate. Second, it allows me to clearly define the nature and role of the audience in this book. Third, it can deal with Balzacq's puzzle of intersubjectivity versus performativity in securitisation theory. The major disadvantage of this reformulation is that it is 'democracy centric' and therefore subject to a different kind of criticism altogether.

A revised securitisation theory

One of the most profound criticisms of securitisation theory to date is that securitisation cannot possibly at the same time operate like an illocutionary speech act and be dependent on acceptance by the audience, because the former denies a meaningful role for the latter.[31] This critique is further strengthened by the fact that if securitisation operated like an illocution, then the distinction, much emphasised by the Copenhagen School, between a 'securitising move' and a complete 'securitisation' would be redundant, as the saying itself (the securitising move) would be the complete securitisation and vice versa.

The distinction between these two events, however, is upheld by the *perlocutionary* speech act. This speech act refers to the performance of a speech act that gets someone else to do something. As such it entails not only a minimum of two actors (the enunciator A and the executor B), but also *two events* – A's utterance of the speech act (securitising move) and B's acting upon it/execution of it, which would make it a complete securitisation. Moreover, and as Balzacq has powerfully argued, *only* the use of the perlocutionary speech act can account for the role of the audience in securitisation theory, which is why he suggests that the perlocutionary speech act captures the logic of securitisation according to the Copenhagen School better than the illocutionary speech act does.[32] Considering, however, that I earlier dismissed the audience as a normative entity derived from Wæver's view of what politics ought to be, securitisation is not accurately captured by the perlocutionary speech act either.

Instead of abandoning speech act theory altogether, another option presents itself, one that upholds the distinction between a securitising

---

[31] Balzacq, 'The Three Faces of Securitization', pp. 175ff.; Stritzel, 'Towards a Theory of Securitization', p. 363.

[32] Balzacq, 'The Three Faces of Securitization', pp. 175ff.

move and a securitisation whilst refraining from theorising a role for the audience in the process of securitisation. Instead of holding that 'the process of securitization is what in language theory is called a speech act',[33] this option holds that only the securitising move is the illocutionary speech act part of securitisation. It holds further that securitisation consists of two events: (1) the securitising move and (2) security practice. Impetus for this formulation is drawn from an early text by Wæver in which the concept of security belonged to the state and in which the audience had no place in securitisation theory. Thus in 1989 Wæver defined the relationship between security and the illocutionary speech act as follows:

What is the illocutionary act in relation to security? It is to define the particular case as one belonging to a specific category ('security') where the state tends to use all available means to combat it. *It is partly a threat but also a kind of promise since more is staked on the particular issue.* The sovereign 'himself' (the regime) is potentially put into question. [T]he *illocutionary force is just a question of succeeding in establishing*: This question is hereby turned into a test-case. It is an open question whether the state then fails – with possible effects for sovereignty or the social order – or it succeeds in blocking developments in case. *The [speech] act in itself just serves to raise the stakes to a principled level.*[34]

This passage suggests that by *speaking* security the securitising actor does not so much 'do security' as later formulations seem to imply, but rather, by employing a 'specific rhetorical structure (survival, priority of action "because the problem is not handled now it will be too late, and we will not exist to remedy our failure")',[35] the securitising actor issues at once a warning to whoever is at the source of the threat and/ or a promise to whoever the actor seeks to protect. 'Warning' and 'promising' are thus 'what is being done' when *performing* the speech act – not security or indeed securitisation. A securitisation exists only at the point when (1) the existential threat justification (the securitising move) is complemented by (2) a change in relevant behaviour by the

---

[33] Buzan *et al.*, *Security: A New Framework for Analysis*, p. 26.
[34] Ole Wæver, 'Security, the Speech Act: Analysing the Politics of a Word', unpublished paper, presented at the Research Training Seminar, Sostrup Manor, 1989, revised, Jerusalem/Tel Aviv 25–26 June 1989, pp. 42–3 (emphasis added).
[35] Buzan *et al.*, *Security*, p. 26.

relevant agent that is justified by this agent with reference to the declared threat (security practice).[36]

My revision thus does two things at once. First, it sets the bar for the existence of securitisation significantly higher than does the Copenhagen School, for whom an issue is securitised simply when an audience accepts the existential threat justification, without any change of behaviour being required. Second, I lower the bar for the success of securitisation. For the Copenhagen School a securitisation is successful when (1) there is an existential threat justification, (2) there are emergency measures, and (3) there is an alteration to interunit relations by breaking free of rules.[37] In my formulation, a securitisation is successful simply when it is brought into existence. The Copenhagen School's reason for setting the bar for the success of securitisation comparatively high is that they want to be able to 'sort the important cases [of securitisation] from the less important ones'.[38] They declare that 'security analysis is interested mainly in studying successful instances of securitization'.[39] If, however, we focus only on the most dramatic securitisations we are at risk of losing sight of something very important, namely the intentions of securitising actors. It is precisely from those cases of securitisation which, according to the Copenhagen School, are unsuccessful, that we can gain insights into why actors securitise, insights that are not easily obtainable from successful securitisations.[40] The crucial point here is

[36] One question that arises in this context is, why does my version of securitisation theory proposed here insist on the speech act element at all? And related to that, could securitisation not consist of security practice alone? Whilst it is indeed the case that the speech act might not in any strict sense be necessary for an issue to be or to become a security issue (formerly top secret archived materials often reveal surprising insights on what sort of issues were considered security issues for any given government without anyone knowing), it is also the case that stating one's objectives and priorities is an intrinsic part of governmental accountability. Moreover, considering that my revised securitisation theory locates 'the audience' with the electorate, it is not strictly true that no one needs to be 'convinced' of the necessity and rightfulness of the securitisation. For the securitisation analyst in turn, language – the securitising move – is important because it helps her to identify who the securitising actor is.

[37] Buzan *et al.*, *Security*, p. 26.

[38] *Ibid.* p. 25.

[39] *Ibid.* p. 39.

[40] It should be noted in this context that Wæver is perfectly aware that studies of these kinds of securitisations are useful. He argues: 'there is a need for more work on de-securitisation and failed acts of securitisation, and finally partial

that even if a securitisation is not followed by the actions which constitute the School's criteria for success, the securitiser still has reasons why they securitised. This is important, because why would anyone opt for securitisation, only to stop short of doing that which securitisation explicitly and uniquely allows: rule-breaking and the introduction of emergency measures? Indeed, stopping short of doing this begs questions regarding the sincerity of securitising actors regarding their existential threat justification. This is not to say that the architects of successful securitisations are always and necessarily sincere in their intentions (as a case in point consider the many speculations over the *real* reasons behind the 2003 US/British-led Iraq invasion), but only that in such cases it is less obvious when what is done in response to the security problem does not match the rhetoric of the existential threat justification. The Copenhagen School, given their sole focus upon successful securitisations, are a good example of this. They do not ask any questions regarding the sincerity of a securitising actor. In the School's writings it is simply assumed that actors revert to securitisation in order to secure the referent object they previously identified as existentially threatened in its existence, hence that the beneficiary of securitisation is the referent object identified by the securitising actor.[41] As such they ignore the fact that in order for a securitisation to be what I will henceforth call 'referent object benefiting', two conditions need to be fulfilled: (a) that the

securitisation could be worth exploring. This is probably increasingly relevant in Western Europe. For instance the Danish involvement in the security of the Baltic States is not about existential threats to Denmark, but the issue has been given some security rationale. Still, this is certainly not enough to lift it out of normal politics; quite the contrary, much of this cooperation is rather de-politicised and "technical" due to its low urgency. [. . .] The currently quite absolutist concept of "securitised or not", might be differentiated through empirical studies of mixed and partial situations' (Wæver, 'Securitisation: Taking Stock', p. 26).

[41] This is perhaps all the more surprising considering that Wæver and the Copenhagen School clearly realise that securitisation can be abused by power holders for self-serving purposes. Wæver writes, 'power holders can always use the instrument of securitization; by definition a problem is a security problem when they declare it to be. [. . .] It is basically an unsolvable "problem" that those administering this "order principle" [securitisation] can easily use it for specific self-serving purposes' (*Concepts of Security*, p. 221). Knowing this, or simply being aware of this, however, is notably different from being able to prove that this is the case and when it is the case. With my revised securitisation theory the analyst can do both.

securitising actor seriously intends to secure the referent object identified; and (b) that the securitising actor acts to alleviate the insecurity
he himself identified. To put the same point another way, referent
object benefiting securitisation assumes the fulfilment of Austin's
sincerity and accordance conditions. Given that, if unfulfilled, these
two conditions do not render the speech act void, but merely
'unhappy' (compare with Chapter 1), it is conceivable that 'unhappy'
or 'infelicitous' securitisations exist.

I propose that the telling characteristic of such infelicitous securitisations is a discrepancy between what securitising actors say and what
they do – between the securitising move and security practice. In order
to test for discrepancy, we need to focus on whether a securitisation
was *consistent* on its own terms. That is, whether or not relevant
behavioural change of relevant actors is or was consistent with the
threats the securitising actors themselves identified. If this is the case,
then we are dealing with a *referent object benefiting securitisation*. If,
however, there exists a considerable and otherwise inexplicable gap
between security practice and the existential threat justification then
we can conclude that the securitisation benefits someone or something
other than the stated referent object. The obvious candidate for such
a beneficiary is the securitising actor, because all securitisations
bestow agents involved in security provision with a raison d'être.
The difference between such an *agent benefiting securitisation* and a
referent object benefiting securitisation is that, whilst in the former the
primary beneficiary of securitisation is the securitising actor, in
the latter this is merely a side-product of securitisation. The answer
to the question 'why did a given actor securitise?' is thus inseparable
from the answer to the question 'who or what is benefited by the
securitisation?'

Distinguishing between different types of securitisation according to
the beneficiary is important beyond allowing insights into intentions
of securitising actors; it suggests that not all securitisations are morally
equal. It holds open the possibility that, depending on who or what
benefits from any given securitisation, it can be either morally right
or morally wrong. Notably, this is contrary to the Copenhagen School
who maintain that *ceteris paribus* securitisations are necessarily morally wrong whilst they hold that desecuritisations are necessarily morally right. For the latter to be true, however, desecuritisation would have
to lead always and necessarily to the same outcome, something the

Copenhagen School quite clearly believe. They understand desecuritisation as the process whereby issues are moved out of 'the threat–defence sequence and into the ordinary public sphere' where they can be dealt with in accordance with the rules of the (democratic) political system.[42] For the Copenhagen School thus, desecuritisation *always* leads to politicisation. Wæver defines politicisation as the state of affairs whereby

> an issue is part of public policy, requiring government decision and resource allocations or some other form of communal governance. [It] means to make an issue appear to be open, a matter of choice, something that is decided upon and that therefore entails responsibility, in contrast to issues that either could not have been different (laws of nature) or should not be put under political control (e.g. a free economy, the private sphere, and matters for expert decision).[43]

This definition of politicisation is, however, but one of three possible definitions in existence; the concept can also refer to the state of affairs where virtually everything is political or, more narrowly, to that which carries official political authority only. Importantly, the wider the conceptual boundaries, the more self-evident is the claim that desecuritisation leads to politicisation. Ultimately, if everything is political, then anything that is not securitised is by definition politicised. On the other hand, this logic also means that when the political is defined narrowly as resting with political authority only, then desecuritisation does *not* automatically lead to politicisation. Thus it is all too feasible that, following desecuritisation, an issue may no longer be part of the political agenda for those in power, even if it is still a concern for other actors. From within a narrow position on politicisation we can therefore say that desecuritisation can lead to politicisation as well as *depoliticisation*.

Earlier on in this chapter I rejected the idea that securitisation is an intersubjective process between a securitising actor and a designated audience. In its place I argued that in questions of national security the power to securitise rests with official political authority only. Considering that the act of securitisation itself is still part of politicisation, this book is informed by the narrower reading of politicisation and the political. Given this, it is possible to identify two types of

---

[42] Buzan *et al.*, *Security*, p. 29.
[43] Wæver, 'Securitisation: Taking Stock', pp. 10, 12.

desecuritisation. The first of these is 'desecuritisation as politicisation', which refers to the process whereby, even after desecuritisation, the issue in question remains on the political agenda of those with official political authority. The second is 'desecuritisation as depoliticisation', whereby the issue disappears from the political agenda of the latter. In this book, therefore, even if other actors within the wider political field (for example, NGOs, academics, opposition party politicians, mid-level staffers) continue to work on some of the same issues after desecuritisation, such activity does not *constitute* politicisation. Only if the government itself continues to work on these issues do we have a case of 'desecuritisation as politicisation'.

The distinction between 'desecuritisation as depoliticisation' and 'desecuritisation as politicisation' is important. In keeping with my earlier distinction between different securitisations, it suggests that not all desecuritisations are morally equal. More precisely, it suggests that there exist morally wrong desecuritisations as well as morally right desecuritisations. Taken together, these distinctions enable me to think cogently about the moral evaluation of desecuritisation and securitisation in the environmental sector of security. This analysis can be found in Chapter 6. For now, two things are important to note in this context.

First, although the realisation that securitising actors may have different intentions from those stated gives the impetus for the moral evaluation of environmental security, it is vital to note that I will not evaluate securitisations in the environmental sector of security in terms of the intentions securitising actors have, but rather in terms of the various consequences produced by the various environmental security policies. This is an important distinction because the logic of consequentialism, whereby the right action is the one that maximises the best consequences, denies a role for intentions. Strictly speaking, however, this is true only when the analyst is concerned with the *actual* consequences of an action, not when he is concerned with the *probable* (also known as foreseeable or expected) consequences of an action. Actual-consequences consequentialism does not distinguish between, on the one hand, acts made with a view to maximising utility that go wrong, and, on the other, acts which, although they are made with a view to not maximising utility, end up 'going wrong' and doing just that. Contrary to this, probable-consequences consequentialism holds that: 'The right action is not the action that results (or would result) in the most happiness, but the action whose outcome has, *on available evidence*, the greatest expected happiness (or greatest expected utility)'.[44] In other words, probable-consequences consequentialism takes account of whether or not an actor *intended* to maximise utility in any given situation. I make use of this latter type of consequentialism only. Unlike actual-consequences consequentialism it allows me to offer, in addition to simple evaluation of security policies that have already taken place, a forward-looking theory of deliberation.[45]

Second, it was argued above that Wæver and the Copenhagen School endorse desecuritisation as a way of escaping 'the normative dilemma of speaking and writing security', whereby the securitising actor, who believes in the performative power of language, inescapably assumes the role of securitiser every time she writes about a security issue. The securitisation is successful when the problem raised becomes recognised as a security problem in the academic literature, and/or in the wider policy-making discourse. Under my revised securitisation theory the latter is not possible, because my revision denies the securitising force of language. In writing about a particular securitisation, the securitisation analyst using my theory issues, at most, a warning (she will hardly be in the position to make a promise for protection); even if the issue is subsequently considered in the same way by other security analysts, this does not constitute a securitisation. The issue would be securitised only if an actor with sufficient capabilities would recognise this warning, echo it, and change its own behaviour in response to it. Immunity to the logic of the normative dilemma, however, does not mean that we can simply neglect normative theorising altogether. After all, it is still the case that some security policies are morally better than others and that we as security analysts act responsibly only if we help potential securitisers to distinguish right from wrong.

Conclusion

This chapter proposed a revision of securitisation theory which allows insights into the intentions of securitising actors, and which paves the

---

[44] William H. Shaw, *Contemporary Ethics: Taking Account of Utilitarianism* (Oxford: Blackwell Publishers, 1999), p. 29 (emphasis added).
[45] Marcus G. Singer, 'Actual Consequence Utilitarianism', *Mind* 86 (1977), pp. 72–3.

way for a normative evaluation of securitisation and desecuritisation in the environmental sector of security. It has been shown that although the Copenhagen School touches on both of these issues, they fail as well as refuse to offer a proper account of either. Their refusal to theorise intentionality rests on the failure to distinguish between the philosophically distinct concepts of 'intentions' and 'motives'. In turn, although they consider the moral evaluation of security policies to be important, the School never goes beyond stating that desecuritisation is morally right and securitisation morally wrong. The reason for this myopia lies with what they anticipate as being the consequences of securitisation and desecuritisation, namely de-democratisation and politicisation respectively. In other words, their inability to conceive of alternative consequences of either securitisation or desecuritisation leaves their moral evaluation necessarily incomplete and one-sided.

The benefits of my revised securitisation theory, however, do not stop with the theorisation of intentions and the possibility of moral evaluation in the environmental sector. Instead, the securitisation analyst informed by my framework can offer a much clearer and more detailed picture of the security policy in question than the traditional securitisation analyst. This is because an analyst informed by my framework is *only* restricted to the use of discourse analysis and publicly available texts in the realm of the securitising move, when he examines the existential threat justification. When studying security practice, however, he is free to use research methods other than discourse analysis – for example, interviews, as well as different texts to those that are publicly available. As part of my analysis of US environmental security under the Clinton and Bush administrations, for example, I draw widely on interviews conducted with almost all the main actors involved in these policies. In addition, I make extensive use of documents that were not publicly available but that were hugely relevant for the security policy (for example documents that defined the meaning of environmental security in the US Department of Defense). Being able to make use of these documents and interviews helped me to provide a much better analysis than would have been possible from discourse analysis alone.

3 | *The rise of US environmental security*

Introduction

The theoretical framework of revised securitisation theory developed in the preceding chapter suggests that the securitisation analyst needs to begin by focusing on the securitising move. To reiterate, the securitising move is the illocutionary speech act part of the securitisation (complete only with security practice) whereby the securitising actor *does something* by 'speaking security' in so far as she issues a warning to whoever is doing the threatening and/or issues a promise to those she seeks to protect. The overriding task of this chapter is to uncover the nature of the securitising move in the case of the first Clinton administration's environmental security policy. Unless otherwise stated, in this chapter 'the Clinton administration' refers to Clinton's first term in office from 1993 to 1997. Since environmental threats are in Gywn Prins' apt phrase 'threats without enemies', the securitising move is likely to have consisted of a promise to protect someone or something only.[1]

Although this chapter aims to trace the history of the rhetorical acknowledgement of environmental security issues (the securitising move) only on the part of the Clinton administration (which means that at this stage I am not claiming that the US environmental security strategy constituted a case of successful securitisation), I cannot avoid touching on at least some of the reasons why the environment was made an issue of national security in the process. This is the case because there is considerable overlap between the reasons for the securitisation and the origins of US environmental security.

---

[1] Gwyn Prins, *Threats without Enemies: Facing Environmental Insecurity* (London: Earthscan Publications, 1993). Note though that this dynamic might change with the increasing importance of climate security, where intentionality in form of *intentional neglect* to do something against climate change becomes an issue.

Agency and structure in US national security

In order to understand US national security policy it is first of all
necessary to understand that this policy is not comprised of the efforts
of a single institution or indeed of a single individual alone. Instead it
stems from the efforts of a complex institutional network, the national
security establishment, that commonly engages the President, the
National Security Advisor, the National Security Council (NSC), most
importantly constituted by the Secretary of Defense, the Secretary of
State and the Vice President, and, to a lesser extent, the Central Intelligence Agency (CIA) and Congress. In the case of US environmental
security the national security establishment was the *securitising actor*.
Although, as will be shown, some of the agencies comprising the
national security establishment ultimately have had more to do with
the provision of environmental security than others (above all the
Department of Defense, DOD), it makes no sense to split the executive
into small and detached units and compare and contrast what each
individual agency said with what they did. That the security establishment has to be taken as one is vividly clear from the fact that the
environmental security policy pursued by the DOD was in fact
thought up and endorsed by members of Congress and soon to be
Vice President Albert Gore.

Like all policy-making areas in the US system, the structure of the
national security establishment resembles the checks and balances at
work in the entire system of governance. To give a very basic idea
of what this means in practice, consider that: first, in all questions
of national security, the President relies on the NSC merely for
advice, which he is free to ignore, yet he is constrained by the
Congress's control over the legislature as well as over the budget
in what he can actually achieve. Second, the National Security
Advisor and the President influence one another, with the National
Security Advisor dependent on a good relationship with the President, but, at the same time, with the National Security Advisor in a
very powerful position to influence the President. Third, the internal
structure of the NSC balances the different preferences prevailing
in the Department of State (DOS), in the DOD and with the
National Security Advisor. Finally, the Congress, despite its control
over the budget, cannot simply allocate money for national security
policies however it sees fit, as many funds are tied up in the

bureaucracies of existing institutions, leaving relatively little room for manoeuvre.[2]

Despite these seemingly strong structural constraints within the US national security establishment, there is a clear role for agency in the making of US national security, because agents (and here it should be noted that in the context of national security it is mostly agencies led by individuals that exercise agency) can shape the agenda by actively creating a need for their own existence, in the process giving a direction to structure (i.e. what is funded and what not). In this book, then, structure and agency in the making of US environmental security are treated as equally important and mutually constitutive.

The work of the national security establishment finds its formal expression in the US National Security Strategy (NSS). The US NSS is the highest official document that states US foreign and security policy.[3] It has three overall objectives, to: '(1.) Furnish a historical perspective to past strategic structure. (2.) Delineate the interest of the United States. (3.) Analyze the threat and objectives of the United States, and the means to achieve these objectives'.[4] The NSS offers a unique strategy. As it is given by the President, it surpasses the division of power between the Congress and the President, and enables the US government to speak with unity. More so, the focus upon the President is seen as particularly useful in the event of crisis. In the articulation of the NSS, the President acts as the highest state representative, policy-maker, chief of staff, and combines security issues that represent his and his party's interests with advice from the National Security Council, the Department of Defense, the Department of State, the Central Intelligence Agency, and the Office of Management and Budget.

---

[2] Sam C. Sarkesian, John Allen Williams and Stephen J. Cimbala, *US National Security: Policy-makers, Processes and Politics*, third edition (Boulder: Lynne Rienner, 2002), pp. 97ff.

[3] The United States had no written National Security Strategy until 1986, when Congress passed the so-called Goldwater–Nichols Act. Though not its primary purpose, this act requires 'that every US administration should produce (approximately once per year) an overview of its national security strategy'. Roland Dannreuther and John Peterson, 'Introduction: Security Strategy as Doctrine', in Roland Dannreuther and John Peterson (eds.), *Security Strategy and Transatlantic Relations* (Abingdon: Routledge, 2006), p. 7.

[4] C. J. Fairchild, 'Does our Nation's Security Strategy Address the Real Threats?' (1989), at www.globalsecurity.org/military/library/report/1989/FCJ.htm [6/2003].

According to the United States Commission on National Security for the 21st Century, also known as the Hart–Rudman Commission:[5] 'American national security strategy must find its anchor in US national interests, interests that must be both protected and advanced for the fundamental well being of American society'.[6] The national interest is commonly defined by what is most important to the state, with state survival usually claiming supremacy. Jack Goldstone offers a reasonable analysis of what counts as a national security issue and what qualifies as such an interest:

A 'national security' issue is any trend or event that (1) threatens the very survival of the nation; and/or (2) threatens to drastically reduce the welfare of the nation in a fashion that requires a centrally coordinated national mobilization of resources to mitigate or reverse. While this seems common sense, it is clear from this definition that not any threat or diminution of welfare constitutes a national security threat; what does constitute such a threat is a matter of perception, judgement, and degree – and in a democracy, a legitimate subject of national debate.[7]

The Hart–Rudman Commission distinguished national interests at three levels: 1. *Survival interests*, without which America would cease to exist as we know it. 2. *Critical interests*, which are causally one step removed from survival issues. 3. *Significant interests*, which importantly affect the global environment in which the United States must act.[8]

As this brief overview shows, the making of US national security policy involves a complex policy-making process deeply anchored in the constitutional division of power within the US system. The international security environment, presidential leadership, public opinion, and intelligence gathering, amongst other factors, all play a role in what qualifies as a national interest and by extension as a national security issue.

---

[5] This bipartisan commission takes its unofficial name from the former US Senators Gary Hart (D) and Warren Rudman (R), who oversaw the commission's work.

[6] Phase II report on a US National Security Strategy for the 21st Century, 'Seeking a National Strategy: A Concert for Preserving Security and for Promoting Freedom' (Washington DC: US Commission on National Security 21st Century, 2000), p. 7.

[7] Jack Goldstone, 'Debate', *Environmental Change and Security Project Report* (Washington DC: The Woodrow Wilson Center, 1996), p. 66.

[8] Phase II Report, p. 7.

US national security after the Cold War

With the end of the Cold War, and thus the removal of the Soviet threat, US national security policy was thrown into a state of disarray, leaving the country *without* an easily definable national security policy. This was a time in which so-called new threats to security, one of which was the potential security implications of environmental degradation, quickly rose to policy-makers' attention, and moreover, *could* rise to attention. The first time 'environmental security' was mentioned in the NSS was in 1991, when George H. W. Bush, in outlining the interests and objectives of the United States for the 1990s, declared that: 'The United States seeks, whenever possible in concert with its allies, to [...] achieve cooperative international solutions to key environmental challenges, assuring the sustainability and environmental security of the planet as well as growth and opportunity for all'.[9] With regard to the argument made earlier, two things can be concluded: first, by 1991, the stability of the environment had become a national interest; and second, environmental problems were seen as having implications for US national security. Given that, in spite of occasionally forbidding conditions, the environment was never openly cast in such a way before 1991, it must be asked: where did this concern for the environment come from? This is an important question, particularly because its origins cannot be traced to a single high profile (environmental) disaster, such as 9/11 for the 'war on terror'.[10] In trying to answer this question, the single most important issue to consider is that the incorporation of the environment (along with other non-traditional security issues) into the NSS took place right after the end of the Cold War. This is no coincidence. As the remainder of this section will show, the end of the Cold War played a pivotal role in the expansion of national security to include non-traditional issues, one of which was the environment.

In searching for explanations for the changed nature of US national security following the end of the Cold War a number of security

---

[9] *National Security Strategy of the United States*, August 1991, at www.fas.org/man/docs/918015-nss.htm [6/2004].

[10] Chernobyl did play a part in the discursive link between 'environment' and 'security' overall, therefore it naturally influenced the discourse in the US; however, this influence was nowhere near the scale of 9/11's meaning for national security.

scholars have suggested that the new security studies (here referring to both real-world policy-making agendas as well as the academic sub-discipline) resemble security studies prior to the onset of the Cold War.[11] David Baldwin famously argued that the Cold War constituted an unnatural state of 'militarized American security policy', whilst 'it also militarized the study of security'.[12] He further suggested that post-Cold War security policy resembled 'some of the modes of thought, policy concerns, concepts of security, and discussions of statecraft' of the world of 1945–1955, where non-militaristic aspects of war, i.e. legal, moral, biological, philosophical etc., were issues of concern.[13]

If it is true that the Cold War and its military focus constituted not the norm but the exception for (US) security policy thinking, then it must follow that other non-military security issues were present at all times, if, however, 'overlaid' by the Cold War.[14] Hence for the analysis of this chapter it must be asked: was the environment, or rather its ill-functioning, already acknowledged as a potential security issue during the Cold War? As a result of overlay it is naturally difficult to answer this question satisfactorily. However, a number of studies suggest a tentative positive answer to this question. As part of a comprehensive study on past presidents and the environment,

---

[11] See, for example, Franklyn Griffiths, 'Environment in the US Security Debate: The Case of the Missing Arctic Waters', *Environmental Change and Security Project Report* (Washington DC: The Woodrow Wilson Center, 1997); Richard Smoke, *National Security and the Nuclear Dilemma: An Introduction to the American Experience in the Cold War*, third edition (New York: McGraw-Hill, 1993); Braden Allenby, 'New Priorities in US Foreign Policy: Defining and Implementing Environmental Security', in Paul Harris (ed.), *The Environment, International Relations and US Foreign Policy* (Washington DC: Georgetown University Press, 2001), pp. 45–67; Mary Margaret Evans, John W. Mentz, Robert Chandler and Stephanie L. Eubanks, 'The Changing Definition of National Security', in Miriam R. Lowi and Brian R. Shaw (eds.), *Environment and Security: Discourses and Practices* (New York: St Martin's Press, 2002), pp. 11–31.

[12] David Baldwin, 'Security Studies and the End of the Cold War', *World Politics* 48 (1995), p. 125.

[13] *Ibid.* p. 141.

[14] The term 'overlay' here is an analogy to Barry Buzan's concept of overlay, whereby 'one or more external powers move directly into the local [security] complex with the effect of suppressing the indigenous security dynamic'. Barry Buzan, *People, States and Fear: An Agenda for International Security Studies in the Post-Cold War Era*, second edition (Hempel Hempstead: Harvester Wheatsheaf, 1991), pp. 219–20.

Raymond Tatalovich and Mark J. Wattier show that the language of security with regard to the environment goes as least as far back as the 1970s. In 1972 the Democratic Party voiced that:

Our environment is most threatened when the natural balance of an area's ecology is drastically altered for the sole purpose of profits. Such practices as 'clear cut' logging, strip mining, the indiscriminate destruction of whole species, creation of select ocean crops at the expense of other species and the unregulated use of persistent pesticides cannot be justified when they threaten our ability to maintain a stable environment.[15]

What is more, the actual term 'environmental security' was already in use in the 1970s, when on the 1976 Democratic Party political platform Democratic leaders attacked the Republican party in arguing that: 'The Democratic "comprehensive program" to achieve "*basic environmental security*" was "thwarted by an administration committed only to unfounded allegations that economic growth and environmental protection are incompatible"'.[16] How environmental security was defined here is not clear and at this stage of the analysis does not matter. What matters here, however, is that the language of security and environment was already interlinked during the Cold War, suggesting the overlaid presence of environment-related security issues during that era. Indeed, President Reagan's 1988 National Security Strategy included the following paragraph:

Critical shortages of food, a lack of health services, and inabilities to meet other basic needs will keep millions of people, particularly in Africa, in peril. The dangerous depletion or contamination of the natural endowments of some nations – soil, forests, water, air – will add to their environmental health problems, and increasingly to those of the global community [. . .] But all create potential threats to peace and prosperity that are in our national interest, as well as the interests of the affected nations.[17]

---

[15] Raymond Tatalovich and Mark J. Wattier, 'Opinion Leadership: Elections, Campaign, Agenda Setting, and Environmentalism', in Dennis L. Soden (ed.), *The Environmental Presidency* (New York: State University of New York Press, 1999), p. 155.

[16] *Ibid.* (emphasis added).

[17] US National Security Strategy from 1988 cited in Geoffrey D. Dabelko, 'Tactical Victories and Strategic Losses: The Evolution of Environmental Security', unpublished doctoral thesis, Faculty of the Graduate School of the University of Maryland (2003), p. 50.

Furthermore, the fact that many of the programmes which from 1996 onwards made up the bulk of DOD's environmental security directive were in existence long before the language of environmental security was used, clearly points to an awareness of environmental security issues during the Cold War. Similarly, it should be noted that the US DOD participated in NATO-led environmental security efforts from the 1980s onwards, and that the US State Department had 'recognized for over two decades [prior to 1995] the relevance of environmental issues in foreign policy'.[18]

Given all that has been said here, it appears that scholars such as Baldwin are right in suggesting that the focus of US national security was unusually narrow during the Cold War. Moreover, the fact that environmental security concerns were already present before and during the Cold War makes it easier to comprehend why the issue gained prominence after the Cold War, as with the end of military 'overlay' such issues were out in the open and ready to be (re-)evoked. The latter argument has been proposed by David Campbell in his 1992 study of US foreign policy *Writing Security*. As part of this study Campbell argues that states rely on so-called 'discourses of danger' to overcome the crisis of representation; for on what other grounds should a state, in protecting its citizens from a threatening 'other', justify its domineering role, other than on grounds of the protection of life, liberty and property? According to Campbell the 'discourses of danger' evoked in the post-Cold War era were thus not new; rather 'there has always been more than one referent around which danger is crystallized'; thus 'what appears as new is often the emergence of *something previously obscured* by that which has faded away or become less salient'.[19] The environment was one such issue. Campbell argues further that because the state is threatened with 'withering away' were it to end its practices of representation, states constantly pursue discourses of danger, with the ultimate threat to state survival being the *absence* of any such discourse.[20] This argument is, however,

[18] Kenneth Thomas, 'Official Statement, Department of State/Office of Under Secretary for Global Affairs', *Environmental Change and Security Project Report* (Washington DC: The Woodrow Wilson Center, 1995), p. 84.

[19] David Campbell, *Writing Security*, second edition (Minneapolis: The University of Minnesota Press, 1998), p. 171 (emphasis added).

[20] *Ibid.* p. 12.

somewhat extreme and it is not clear how the claim that the US as a country has no other source of identity than a threatening 'other' can be substantiated. Be that as it may, there is something useful in the idea of 'discourses of danger' and in a slightly changed form this concept will play a significant role in this book. Hence, it will be argued that it is not the state that suffers a crisis of representation in the absence of danger, but rather the national security agencies do, for providing security is their sole raison d'être. In other words it is the national security establishment, not the state, that actively pursues discourses of danger.

In summary, the end of the Cold War was important for the rise of environmental security in two related ways. First, concerns about the implications of an ill-functioning environment for national security predated the end of the Cold War and were ready to be re-evoked. And second, in need of new discourses of danger to maintain integrity, security agencies turned to some of the existing but previously overlaid discourses of danger, one of which was the environment. As Geoffrey D. Dabelko suggests, 'the end of the Cold War presented a window of opportunity for big ideas that challenged long-held assumptions of international relations and security in particular'.[21]

Whilst the rise of US environmental security cannot be understood without the end of the Cold War, it further cannot be understood without recognising the status of environmental issues in politics during that time. Hence, coinciding with the end of the Cold War there was significant international focus on general environmental and global issues due to the run-up to the Earth Summit in Rio in 1992. Generally speaking, in the early 1990s, the environment was no longer the minority concern it had been in the 1960s, when Rachel Carson's widely acknowledged work *Silent Spring* (1962) projected environmental issues into mainstream discourse. Notable steps in the establishment of the environment as a majority concern have been the 1972 UN Conference on the Human Environment (Stockholm), the World Conservation Strategy (1980), the Brundtland report (1987), plus an array of environmental conventions and regimes, such as the London Dumping Convention (1972) and, perhaps most famously, the Montreal Protocol (1987).

---

[21] Dabelko, *Tactical Victories and Strategic Losses*, p. 1.

The meaning of US environmental security

Having established that the rise of environmental security was inseparably tied to the end of the Cold War let us now turn to the securitising
move, that is, to the questions of who or what was the referent object of
environmental security for the Clinton administration, and who or what
was at the source of the threat. When the Clinton administration
came into office in January 1993, environmental security had been
part of the academic literature for some time. As early as 1983, for
example, Richard Ullman, at that time Professor of International
Affairs at Princeton University's Woodrow Wilson School of Public and
International Affairs, had been arguing for the incorporation of environmental threats alongside traditional military threats into the national
security framework. In 'Redefining Security' Ullman listed a number of
environmental problems that may potentially lead to security implications for the state, with particular emphasis placed on the US. These
environmental issues were: earthquakes, conflicts over territory and
resources, population growth and resource scarcity, particularly oil. To
avert these security implications Ullman argued for the redefinition of the
threat to national security to comprise 'disturbances and disruptions
ranging from external wars to internal rebellions, from blockades and
boycotts to raw material shortages and devastating "natural" disasters
such as decimating epidemics, catastrophic floods, or massive and pervasive droughts',[22] whereby he defined a threat to national security as 'an
action or sequence of events that (1) threatens drastically and over a
relatively brief span of time to degrade the quality of life for the inhabit-
ants of a state, or (2) threatens significantly to narrow the range of policy
choices available to the government of a state or to private, nongovern-
mental entities (persons, groups, corporations) within the state'.[23]

Writing during the Cold War, Ullman was aware that the dominance
of military threats made it difficult even for willing policy-makers to
advocate the significance of other less pervasive threats such as the
environment. The redefinition of national security, at least in the first
instance, could therefore not be a top-down process, but had to
commence with enhanced 'public education' about the threat potential
of an ill functioning environment.[24] This then, Ullman suggested,

[22] Richard Ullman, 'Redefining Security', *International Security* 8 (1983), p. 133.
[23] *Ibid.* p. 133.
[24] *Ibid.* pp. 152–3.

would lead to the general acceptance of the environment as a security issue, whilst at the same time enabling policy-makers to treat environmental issues as national security issues.

Some of the writings that followed this early publication on the need to redefine national security did provide such 'public education', for both policy-makers and the academic literature. The mostly descriptive works of the prominent environmental scientist Norman Myers and then Vice President of the World Resources Institute Jessica Tuchman Mathews, for instance, sought to enhance awareness of the scale and importance of environmental issues. In using detailed descriptions of the state of the environment, accompanied by bleak predictions about the future, these writers made a case for recognising the pressing urgency of environmental problems and the need for action.[25] Although they were very broad in their take on environmental security – indeed with their focus on the well-being of the individual they can easily be positioned within the human security approach to environmental security (see Chapter 6) – it is telling that they chose to advocate a fairly narrow and state-centric version of environmental security instead. For example, Myers argued that 'national security is no longer about fighting forces and weaponry alone. It relates increasingly to watersheds, croplands, forests, genetic resources, climate, and other factors rarely considered by military experts and political leaders, but that taken together deserve to be viewed as equally crucial to a nation's security as military prowess'.[26] The reason why these authors focused on national security despite being interested in human security was that they realised that their voices were more likely to be heard if they remained within the traditional state-centric reading of security.[27] That they were right to think this becomes clear when considering that, amongst other

[25] See, for example, Norman Myers, *Ultimate Security: The Environmental Basis of Political Stability* (New York: Norton, 1993); Norman Myers and Julian Simon, *Scarcity or Abundance? A Debate about the Environment* (London: W. W. Norton & Company, 1994); Norman Myers, 'Environment and Security', *Foreign Policy* 74 (1989), pp. 23–41; Norman Myers, 'The Environmental Dimension to Security Issues', *The Environmentalist* 6 (1986), pp. 251–7; Jessica Tuchman Mathews, 'Redefining Security', *Foreign Affairs* 68 (1989), pp. 162–77.

[26] Myers, *Ultimate Security*, p. 21.

[27] Dabelko, *Tactical Victories and Strategic Losses*, pp. 4, 45. See also S. Neil MacFarlane and Yuen Foong Kong, *Human Security and the UN: A Critical History* (Indianapolis: Indiana University Press, 2006), p. 233.

things, their writings served as a means for policy-makers to 'explain activities and raise the profile of international environmental concerns', which taken together launched the subsequent environmental security policy in the United States.[28] Rhetorical acknowledgements of the writings by Mathews *et al.* were clearly visible in early speeches by the Clinton administration. The following passages taken from President Clinton's inaugural speech, his 1993 Earth Day speech and his State of the Union address of the same year highlight this well:

Americans deserve better, and in this city today, there are people who want to do better. And so I say to all of us here, let us resolve to reform our politics, so that power and privilege no longer shout down the voice of the people. Let us put aside personal advantage so that we can feel the pain and see the promise of America. [...] To renew America, we must meet challenges abroad as well as at home. There is no longer division between what is foreign and what is domestic – the world economy, the world environment, the world AIDS crisis, the world arms race – they affect us all.[29]

Unless we act, and act now, we face a future where our planet will be home to nine billion people within our lifetime, but its capacity to support and sustain our lives will be very diminished. Unless we act now, we face a future in which the sun may scorch us, not warm us; where the change of season may take a dreadful new meaning; and where our children's children will inherit a planet far less hospitable than the world in which we came of age.[30]

Backed by an effective national defense and a stronger economy, our nation will be prepared to lead a world challenged – as it is everywhere – by ethnic conflict, by the proliferation of weapons of mass destruction, by the global democratic revolution and by challenges to the health of our global environment.[31]

In later speeches, by Clinton and other White House officials, these arguments became ever more developed, with the result that environmental security became laid out in the 1994 and 1995 US National

[28] Geoffrey D. Dabelko and P. J. Simmons, 'Environment and Security: Core Ideas and US Government Initiatives', *SAIS Review* 17 (1997), p. 132.
[29] President William J. Clinton, 'Inaugural Address', 20 January 1993, at www.presidency.ucsb.edu/ws/index.php?pid=46366 [12/2004].
[30] 'President Clinton's Remarks on Earth Day 1993', *Environmental Change and Security Project Report* (Washington DC: The Woodrow Wilson Center, 1995), p. 50.
[31] President William J. Clinton, 'State of the Union', 17 February 1993, at www.presidency.ucsb.edu/ws/index.php?pid=47232 [12/2004].

Security Strategy of Engagement and Enlargement. Of which extracts read as follows:

Protecting our nation's security – our people, our territory and our way of life – is my Administration's foremost mission and constitutional duty. The end of the Cold War fundamentally changed America's security imperatives. [...] The dangers we face today are more diverse. [...] Large scale environmental degradation, exacerbated by rapid population growth, threatens to undermine political stability in many countries as regions. [...] In all cases, the nature of our response must depend on what best serves our own long-term national interests. Such requirements start with our physical defense and economic well-being. They also include environmental security. [...] The range of environmental risks serious enough to jeopardize international stability extends to massive population flight from man-made or natural catastrophes, such as Chernobyl or the East African drought, and to large-scale ecosystem damage caused by industrial pollution, deforestation, loss of biodiversity, ozone depletion, desertification, ocean pollution and ultimately climate change. [...] Rapid population growth in the developing world and unsustainable consumption patterns in industrialized nations are the root of both present and potentially even greater forms of environmental degradation and resource depletion.[32]

Though the extent of these passages may seem excessive they are necessary for our purposes here as they hold vital clues regarding who the Clinton administration identified as being existentially threatened in their existence with regard to 'environment' and 'security', and what, in turn, was responsible for the threat. Hence they suggest that none other than the American people were believed existentially threatened in their existence, whilst environmental degradation of any kind (including global, regional and domestic issues) was identified as the source of the threat. To put this differently yet again, in rhetorically drawing a connection between 'environment' and 'security' the US government made a promise to US citizens to protect them from threats associated with environmental change.

In 1994, with the arrival of the so-called environmental scarcity thesis in policy-making circles, the Clinton administration's environmental security discourse gained yet another dimension. Generally speaking, the environmental scarcity thesis focuses on threats to the

[32] '1994 and 1995 US National Security Strategy of Engagement and Enlargement', extracts reproduced in the *Environmental Change and Security Project Report* (Washington DC: The Woodrow Wilson Center, 1995), pp. 47ff.

security of the state resulting from the linkage between environmental scarcity and violent conflict.[33] It first came to high-ranking policy-makers' attention through the publication of the February 1994 edition of the *Atlantic Monthly*, which contained the article 'The Coming Anarchy', by the prominent journalist Robert Kaplan.[34] In this article Kaplan portrays the environment as a 'hostile power', one that can only be controlled if the environment is made a national security issue. In Kaplan's own words:

It is time to understand 'the environment' for what it is: the national security issue of the early twenty-first century. The political and strategic impact of surging populations, spreading disease, deforestation and soil erosion, water depletion, air pollution and, possibly, rising sea levels in critical,

---

[33] See for example: Thomas F. Homer-Dixon and Val Percival, 'The Case of South Africa', in Paul F. Diehl and Nils Petter Gleditsch (eds.), *Environmental Conflict* (Oxford: Westview Press, 2001), pp. 13–35; Thomas F. Homer-Dixon, *Environment, Scarcity, and Violence* (Princeton: Princeton University Press, 1999); Thomas F. Homer-Dixon, 'Environmental Scarcities and Violent Conflict: Evidence from Cases', *International Security* 19 (1994), pp. 5–40; Thomas F. Homer-Dixon, 'On the Threshold: Environmental Changes as Causes of Acute Conflict', *International Security* 16 (1991), pp. 76–116; Günther Baechler, 'Why Environmental Transformation Causes Violence: A Synthesis', *Environmental Change and Security Project Report* (Washington DC: The Woodrow Wilson Center, 1998), pp. 24–44; Günther Baechler, *Violence through Environmental Discrimination: Causes, Rwanda Arena, and Conflict Model* (Dordrecht: Kluwer Academic Publishers, 1999). Please note that the environmental scarcity thesis is not the only environment-conflict thesis in existence. An alternative would be the so-called honey-pot thesis. This thesis holds that it is not scarcity but instead the abundance of valuable natural resources (such as minerals, gem stones or oil) that makes people fight. In short, that it is greed and not grievance that makes people resort to violence. (Indra De Soysa, 'The Resource Curse: Are Civil Wars Driven by Rapacity or Paucity?' in Mats Berdal and David Malone (eds.), *Greed and Grievance: Economic Agendas in Civil War* (Boulder: Lynne Rienner, 2000), pp. 113–35; Michael T. Klare, *Resource Wars: The New Landscape of Global Conflict* (New York: Henry Holt and Company, 2001); Paul Collier and Anke Hoeffler, 'Greed and Grievance in Civil War', working paper (Oxford: Centre for the Study of African Economies, 2002).

[34] This is not the only time Robert Kaplan's writings have risen to policy-makers' attention. His 1994 *Balkan Ghosts* 'was said, by Richard Holbrooke, then a key figure in the Clinton Administration, to have played a key role in convincing that Administration not to intervene militarily in Bosnia in the middle 1990s. The Administration, it seems, were taken by Kaplan's narration of the conflict based on "ancient hatreds", thus not amenable to resolution from "outside"' (Stuart Croft, *Culture, Crisis and America's War on Terror* (Cambridge University Press, 2006), pp. 56–7.

overcrowded regions [...] will be the core foreign policy challenge from which most others ultimately emanate, arousing the public and uniting assorted interests left over from the Cold War.[35]

This article was studied carefully by President Clinton[36] and it is said to have captured the imagination of Vice President Gore and that of the Under Secretary of State for Global Affairs Timothy E. Wirth. The latter distributed the article to every US embassy worldwide. The February 1994 edition of *The Atlantic Monthly* became one of the best-selling copies of the magazine. As a result of the engagement with Kaplan's article the Clinton administration – particularly the Vice President – became interested in available research into the environmental scarcity thesis, which resulted in the invitation of one of its leading scholars, the Canadian political scientist Thomas Homer-Dixon, to Washington DC in the spring of 1994. The choice of Homer-Dixon clearly was no coincidence, as throughout his article Kaplan makes heavy reference to Homer-Dixon's work and places his own findings within the theoretical framework developed by the former. In the words of one journalist, Homer-Dixon's rise to attention took the following route:

Homer-Dixon co-authored in 1993 a more accessible article on the same topic [more accessible than his 1991 piece 'On the Threshold: Environmental Changes as Causes of Acute Conflict' published in *International Security*] in the *Scientific American*, "Environmental Change and Violent Conflict". That in turn led to a *New York Times* op-ed article, later reprinted in the *International Herald Tribune*. Soon stories trickled down of people reading his original *International Security* article on Air Force One. Copies of the article circulated in the National Security Council, the Pentagon, the State Department, and the Central Intelligence Agency. Al Gore picked up the phone.[37]

Although Homer-Dixon had been in contact with policy-makers in Washington before February 1994 – for example, in 1992 he briefed a

---

[35] Robert Kaplan, 'The Coming Anarchy', in Gearóid Ó Tuathail, Simon Dalby and Paul Routledge (eds.), *The Geopolitics Reader* (London: Routledge, 1998), p. 190.

[36] Author interview with the former Director of CIA (1993–1995) R. James Woolsey, 14 September 2005, Arlington VA.

[37] Mark Kingwell, 'Meet Tad The Doom-meister' *Saturday Night* (September 1995), p. 44. Of course, Gore did not pick up the phone himself, but rather one of his aides did. Nonetheless, from a personal conversation with Gore's national security advisor Leon Fuerth I learned that Gore was very interested in the ideas of Homer-Dixon, with Fuerth himself more sceptical of the entire link.

national security meeting under the auspices of P. J. Simmons, who later went on to direct the Environmental Change and Security Program, created within the Woodrow Wilson Center for Scholars in Washington DC – it was not till *after* the publication of Robert Kaplan's 'The Coming Anarchy' that Homer-Dixon was contacted by the Vice President's staff.[38] During 1994, Homer-Dixon briefed the Vice President twice, once in late April and again in August. The first meeting took place in the fairly intimate setting of a dinner at Gore's Washington residence. This meeting was attended by about ten people, among them Gore's immediate assistants, his wife Tipper Gore, Gore's National Security Advisor Leon Fuerth, Under Secretary of State for Global Affairs Timothy E. Wirth, Deputy Under Secretary of Defense for Environmental Security Sherri W. Goodman and Director of the United States Agency for International Development Brian J. Atwood. The subsequent meeting in August that year was of a different nature and focused specifically on the environmental problems in China. Homer-Dixon, together with his colleagues Jack Goldstone and Vaclav Smil, briefed some thirty to forty administration staff, among them the Administrator of the Environmental Protection Agency Carol Browner, the Under Secretary of the Treasury for International Affairs Larry Summers and the Director of the CIA R. James Woolsey.

Throughout 1996 Homer-Dixon and his colleagues conducted the most extensive round of briefings in Washington DC, and it was then that they distributed the specially edited *Environmental Scarcity and Violent Conflict: Briefing Book* that provided, in less than sixty pages, a concise overview of the work up to that time by the so-called Toronto Group. The group's main messages, as later summarised by Homer-Dixon in *Environment, Scarcity and Violence* (1999), are that '*environmental scarcity* can contribute to civil violence, including insurgencies and ethnic clashes',[39] and that 'in coming decades the incidence of such violence will probably increase as scarcities of cropland, freshwater and forests worsen in many parts of the developing world'.[40]

In Homer-Dixon's models, environmental scarcity has deleterious social effects and it is the *interaction* of these different variables, rather

---

[38] This and the remaining not cited information draw on a telephone interview between the author and Thomas Homer-Dixon on 20 June 2005.
[39] Homer-Dixon, *Environment, Scarcity, and Violence*, p. 177.
[40] *Ibid.*

than the state of the environment alone, that leads to conflict. In Homer-Dixon's words: 'Scarcity's role in [...] violence, however, is often obscure and indirect. It interacts with political, economic, and other factors to generate harsh social effects that in turn help produce violence'.[41] Homer-Dixon *et al.* identify three types of environmental scarcity:

(1) Supply-induced scarcity is caused by the degradation and depletion of an environmental resource, for example, the erosion of cropland; (2) demand-induced scarcity results from population growth within a region or increased per capita consumption of a resource, either of which heightens the demand for the resource; (3) structural scarcity arises from unequal social distribution of a resource that concentrates it in the hands of a relatively few people while the remaining population suffers from serious shortages.[42]

Based on the finding that environmental scarcity serves as an underlying yet strong cause of intrastate conflict, Homer-Dixon argues that the environment needs to be prioritised in order to protect the existing order of the international system from violent conflict. Although he focuses primarily upon the developing world as being directly affected by environmental conflict, the possibility of spill-over to the developed world is given.

---

[41] Homer-Dixon, *Environment, Scarcity, and Violence*, p. 177. As already stated, in the academic world of environmental security Homer-Dixon's work is contested and one particular point of contention is the Toronto Group's selection of case studies. Critics take issue with the fact that all of the group's case studies were, and still are, chosen in the knowledge that both environmental scarcity and violent conflict have existed in the particular country/region selected for the study, leading these critics to assert that the Toronto Group's work offers nothing but 'anecdotal evidence' and little of novel value (Marc A. Levy, 'Is the Environment a National Security Issue?' *International Security* 20 (1995), pp. 35–62). Homer-Dixon justifies his selection of case studies by arguing that it is not the coexistence of environmental scarcity and violent conflict that is subject to research, but rather the causal links between the two phenomena (Thomas Homer-Dixon, 'Debate between Thomas Homer-Dixon and Marc A. Levy', *Environmental Change and Security Project Report* (Washington DC: The Woodrow Wilson Center, 1996), pp. 49–60. More critical scholars still argue that Homer-Dixon's research remains unsatisfying as his positivist methodology fails to answer the important question: 'What makes people resort to violence?': Jon Barnett, *The Meaning of Environmental Security: Ecological Politics and Policy in the New Security Era* (London: Zed Books, 2001), p. 64.

[42] Homer-Dixon *et al.*, 'The Case of South Africa', p. 14.

Soon the biggest contributors to global environmental problems, developing
countries could become more belligerent, less willing to compromise with
other states, and less capable of controlling their territories in order to
implement measures to reduce environmental damage. If many developing
countries evolve in the direction of extremism, the interests of the North
may be *directly* threatened.[43]

[C]onflicts generated partly by environmental scarcity, although perhaps not
as conspicuous or dramatic as interstate wars, can nonetheless have serious
indirect effects on the international community. The changing nature of the
international system – heightened economic interdependence, easier long-
distance travel, and increased access to arms – makes previously
insignificant regions of interest to policy-makers. Crises in small countries,
such as Haiti, often create serious foreign policy difficulties for developed
countries, and large significant countries [...] are not immune to the severe
stresses environmental scarcity generates.[44]

Because of these dangers, states should advocate strategies to close
what Homer-Dixon calls 'the ingenuity gap', namely the gap between
ever-rising requirements for ingenuity resulting from the pace of
modern life and the capacity to deal with these developments.[45]
Further, states should actively support pre-emptive solutions to the
effects of environmental stress, one of which is the spread of liberal
democracy and the spread of increased trade.[46] Based on the premises
of Doyle's 'democratic peace thesis', Homer-Dixon suggests that
democracies strengthened through economic interdependence do not
go to war with each other – not even over environmental scarcities.

There is little doubt that there was interest in Washington at the time
in the idea that environmental scarcity may lead to violent conflict, and
the language of the environmental-scarcity thesis was readily invoked
by a diverse group of policy-makers. The following extracts from
speeches by the Deputy Under Secretary of Defense for Environmental
Security Sherri W. Goodman, US Secretary of State Warren Christopher
and President Clinton, respectively, make this clear.

The role of environmental degradation and scarcity in causing instability
and conflict is the subject of much debate in the academic community. [...]

[43] Homer-Dixon, 'On the Threshold', p. 113 (emphasis added).
[44] Homer-Dixon, *Environment, Scarcity, and Violence*, p. 180.
[45] Thomas Homer-Dixon, *The Ingenuity Gap* (New York and Toronto: Alfred
     A. Knopf, 2000).
[46] Homer-Dixon, 'On the Threshold', p. 115.

Scarcity of renewable resources such as water, forests, cropland, and fish stocks occur from degradation and depletion of resources, overconsumption and overuse of resources, and/or inequitable distribution of resources. [...] Environmental scarcities can interact with political, economic, social and cultural factors to cause instability and conflict. [...] The multiple effects of environmental scarcity, including large population movements, economic decline, and capture of environmental resources by elites, can weaken the government's capacity to address the demands of citizens. If the state's legitimacy and capacity for coercive force are undermined, the conditions are ripe for instability and violent conflict.[47]

President Clinton and I strongly believe that protecting our environment is a central national interest of the United States. Whether in San Francisco or San Salvador, pollution can take a tremendous toll on human health and the prospects for stability and economic development. Damage to the environment can intensify conflicts between nations over shrinking resources and competing needs. That kind of damage to the environment weakens our global economy by harming agriculture and fisheries, by harming forests and manufacturing.[48]

[W]hen you look at the long-run trends that are going on around the world – you read articles like Robert Kaplan's article in *The Atlantic* a couple of months ago that some say is too dour – if you really look at what's going on, you could visualize a world in which a few million of us live in such opulence we could all be starring on nighttime soaps. And the rest of us look like we're in one of those Mel Gibson 'Road Warrior' movies. I was so gripped by many things that were in that article, and by the more academic treatment of the same subject by Professor Homer-Dixon.[49]

With regard to US environmental security more generally it is important to note that the environmental scarcity thesis opened the way for yet a third environmental security policy – the usage of intelligence capabilities in the monitoring of the environment. Although bolstered by the environmental scarcity thesis, this idea actually dates back to

[47] Sherri W. Goodman, 'The Environment and National Security', Speech at the National Defense University, 8 August 1996 at www.loyola.edu/dept/politics/intel/goodman.html [1/2005].

[48] Warren Christopher, 'Address to the World Business Council in San Salvador, El Salvador' (Washington DC: US Department of State Dispatch, 27 February 1996).

[49] Bill Clinton, 'President Clinton's Remarks to the National Academy of Sciences', *The Environmental Change and Security Project Report* (Washington DC: The Woodrow Wilson Center, 1995), p. 51.

Senator Albert Gore's days at the Armed Services Committee. Gore, who as a young student at Harvard University had a formative experience with regard to the science and implications of global warming,[50] was particularly interested in environmental issues and subsequently established himself as an authority on environmental affairs within Congress. In 1988 Gore had experienced the utility of intelligence capabilities for environmental monitoring first hand during a trip to the North Pole, where he observed the cooperation between ice scientists and the US Navy on measuring the thickness of the icecap.[51] The idea to use Cold War era intelligence capabilities to solve the problems of the post-Cold War world, such as environmental degradation, was extremely popular with the Democrats in the Armed Services Committee, because it ensured continuous funding.[52] Gore – together with the Senators (all Armed Services Committee) Sam Nunn, Jeff Bingaman, Timothy Wirth and James Exon – played a pivotal role in this development. The creation of the Strategic Environmental Research and Development Program (SERDP), introduced to the House on 28 June 1990, launched the use of intelligence capabilities for monitoring environmental issues. During the aforementioned Senate hearing Senator Gore described the merits and utility of SERDP as follows:

- First, it recognizes that the environment is producing political, economic and security effects that are meaningful to the concerns and missions of the Department of Defense.
- Second, it recognizes that there are some activities within the Department of Defense and the defense sector of the Department of Energy that are germane to these environmental problems; for example [...] data gathering and analysis, energy, and environmental remediation.
- Third: it recognizes the value to the Nation of managing these activities in such a way as to derive benefits for civilian environmental priorities; and conversely, of calling on the best civilian environmental technologies for ideas of value to the military.

---

[50] This experience is described in some detail in Gore's 1992 book, *Earth in Balance*; its significance for Gore's environmental dedication, however, was confirmed to me in an interview with Gore's national security advisor Leon Fuerth on 12 September 2005. The experience is further mentioned in Gore's 2006 film *An Inconvenient Truth*.
[51] Albert Gore, *Earth in Balance* (London: Earthscan Publications, 1992), p. 22.
[52] Dabelko, 'Tactical Victories and Strategic Losses', pp. 41–2.

- Fourth, it recognizes that certain, sometimes quite serious, environmental problems have been caused within the Department of Defense itself and need to be remedied quickly.
- Finally, it recognizes the need to organize this approach on a sustained long-term basis funded at serious levels.[53]

SERDP was established on 5 November that year, through Public Law 101-510 (10 United States Code 2901–2904) and the use of intelligence for environmental data gathering subsequently acknowledged in the 1994/1995 NSSs.

As point four of the above shows, at that time the dual use of intelligence capabilities was not the only environmental security policy issue instigated by the Armed Services Committee in general and by Gore in particular. Thus the Armed Services Committee suggested that the DOD should change its hitherto negative environmental record and become a good steward of the environment instead. This was not an entirely novel idea: the base cleanup programme, for example, dates back to 1984 when the Alaskan Senator Ted Stevens, in his powerful position as Chairman of the Appropriations Committee, created the programme, in part because he wanted the removal and cleanup of many leftover World War II sites in his own state.[54] Once this programme was in place, more senators wanted money for similar cleanup activities in their own states, leaving the programme to grow. To be sure, the origins of military environmental stewardship could be said to date as far back as the 1960s, when television and printed press coverage from the Vietnam War delivered powerful visual images into the living rooms of American and worldwide audiences, that left little to the imagination as to the horrors caused by environmental warfare.[55]

---

[53] Congressional Record Senate hearing, Legislative day Monday 11 June 1990 (Washington DC: Senate Records), p. 5.

[54] This said, one should not err and assume that Stevens is a great environmentalist; after all he supports drilling in Alaska (David Firestone, 'Drilling in Alaska, a Priority for Bush Fails in the Senate', *New York Times*, 20 March 2003).

[55] One of the most powerful images that highlight the horrors of chemical warfare during the Vietnam War was that of nine-year-old Kim Phuc, running naked from the napalm strike near Trang Bang on 8 June 1972. The picture is available at http://news.bbc.co.uk/2/hi/asia-pacific/4517597.stm In recent years the incorporation of threat images into the study of the security has become more prolific; this is not lest because of Michael C. Williams, 'Words, Images, Enemies: Securitization and International Politics', *International Studies Quarterly* 47 (2003), pp. 511–31.

Environmental warfare, also known as environmental manipulation, refers to the course of conduct whereby the planned destruction of the environment – be it through herbicides, chemical bombs/agents, concussion bombs, forest fires, as well as deliberate salinisation of arable land or freshwater reservoirs, for example through the breaking of dams – constitutes an important part of the overall military strategy.[56] Environmental destruction during warfare, it could be argued, is a necessary evil, as in war destruction is a necessity. In other words, environmental damage as collateral damage during warfare may be excusable as it may be necessary to defend the state. Whether this argument extends to environmental warfare, however, is debatable; as is environmental destruction during military training exercises and in preparation for war.

Studies that analyse the environmental damage done by the US military during the years of the Cold War tell frightening tales of unrestricted environmental destruction, through the toxic contamination of military bases, surrounding areas, groundwater and rivers, and through munitions testing, manoeuvre damage, loss of nuclear/chemical/biological weapons and so forth, all in an effort to maintain the nuclear stand-off between the superpowers. Such environmental damage is not restricted to the US military alone, its Soviet counterpart faring even worse; environmental concern was simply non-existent (see Chapter 4). By the end of the Cold War, this careless treatment of the natural environment left the DOD, which controls 25 million acres of land in the US alone, with an environmental legacy that will take billions of dollars and potentially decades to clean up. The US Army's Rocky Mountain Arsenal in Colorado, for example, has been dubbed 'the most toxic square mile on earth', with its 'soil and groundwater contaminated with pesticides, mustard gas, nerve agents,

[56] For the impact of the military on the environment see Susan D. Lanier-Graham, *The Ecology of War: Environmental Impacts of Weaponry and Warfare* (New York: Walker and Company, 1993); Arthur H. Westing, 'Environmental Warfare: Manipulating the Environment for Hostile Purposes', *Environmental Change and Security Project* (Washington DC: The Woodrow Wilson Center, 1997), pp. 145ff.; Arthur H. Westing (ed.), *Herbicides in War: The Long-term Ecological and Human Consequences* (London: Taylor and Francis, 1984); Arthur H. Westing, *Warfare in a Fragile Environment: Military Impact on the Human Environment* (London: Taylor and Francis, 1980); Gwyn Prins and Robbie Stamp, *Top Guns and Toxic Whales: The Environment and Global Security* (London: Earthscan Publications, 1991).

heavy metals, and incendiaries'.[57] According to Army officials, the estimated cost of cleanup for this particular base alone could be as high as $10–20 billion.[58]

It is fair to say that, until the end of the Cold War, public interest in the military's environmental conduct was a minor issue, seen by most as a necessary evil of the ongoing 'war effort'. With the Cold War over, however, public awareness of the environmental damage done by the military soared to previously unknown heights, leading to public condemnation of detrimental defence activities at home.[59] This public outcry concerned nuclear safety issues at the Department of Energy (DOE) as opposed to the facilities of the DOD directly. In the late 1980s almost all of the nuclear production facilities failed to comply with environmental health standards, leading to a three-month long condemnation campaign in the *New York Times*, spearheaded by journalist Matthew L. Wald. As a result of these developments the George H. W. Bush administration was forced to change its attitude towards environmental stewardship of the military, with the Secretary of Defense Cheney embracing the new spirit. The author and journalist Seth Shulman describes the prevailing mood at the time as follows:

Cheney's decision came only after much prompting. He issues the memo on the eve of the twentieth anniversary of Earth Day, against the backdrop that could only be constructed as a fever pitch of national concern about environmental issues. [...] The same year, Cheney and his staff have watched as forceful newspaper reports about the abysmal state of the government's nuclear production facilities prompted a public and governmental outcry that resulted in a virtual shutdown of the Department of Energy's nuclear production operations. *Many officials undoubtedly feared that to some degree the same could happen to the Department of Defense.*[60]

At a hastily put together conference in 1989, Cheney stated, 'This Administration wants the United States to be the world leader in addressing environmental problems, and I want the Department of

---

[57] Center for Defense Information Washington DC, 'The Military and the Environment', *The Defense Monitor* 23(9) (1994), p. 4.

[58] *Ibid.* p. 4.

[59] Odelia Funke, 'Environmental Dimensions of National Security', in Jyrki Käkönen (ed.), *Green Security or Militarized Environment* (Aldershot: Dartmouth, 1994), pp. 65ff.

[60] Seth Shulman, *The Threat at Home: Confronting the Toxic Legacy of the US Military* (Boston: Beacon Press, 1992), p. 117 (emphasis added).

Defense to be the Federal leader in agency environmental compliance and protection'.[61]

As a result of the public outcry and of the prevailing problems with the nuclear weapons complex, members of Congress – specifically the Senate's Governmental Affairs and the Armed Services Committee – became increasingly interested in the treatment of the environment by the DOE and later by the military. Gore together with Senator Sam Nunn, at the time the chairman of the Armed Services Committee, quickly became the driving force behind the military's environmental security efforts. Together they evoked the idea that the mistreatment of the environment by the DOD leads to environmental security threats, conditions which affect not only human health and safety on and around military bases, but also conditions that impair DOD's ability to prepare for or carry out the NSS, or create instabilities that can threaten US national security. As a result of this new understanding, Congress allocated more resources to environmental cleanup at military bases, and with the Federal Facilities Compliance Act (1992) new legislation was passed. Whilst this act was but one in a long list of environmental statutory provisions it is of particular importance.[62] It enabled the Environmental Protection Agency (EPA) to issue fines, with individual base commanders liable to go to prison in case of environmental neglect or in case of non-compliance with existing environmental legislation. The Federal Facilities Compliance Act gave new teeth to old laws, making compliance a necessity.

Conclusion

This chapter has analysed the rise of environmental security in the US context and the nature of the securitising move. The analysis began by identifying the national security establishment as the *securitising actor* in the case of US environmental security. Consequently, the

---

[61] Cheney quoted *ibid*. p. 116.

[62] From 1970 onwards the US government passed a number of significant environmental laws all of which applied to DOD facilities. Among others these were: The Occupational Safety and Health Act (1970); The Clean Air Act (1970); The Endangered Species Act (1973); The Resource Conservation and Recovery Act (1976); The Clean Water Act (1977); Comprehensive Environmental Response, Compensation, and Liability Act (1980); The Pollution Prevention Act (1990).

national security strategy, in which the work of the national security establishment finds its formal expression, was considered the most significant publicly available document from which to trace the securitising move. It was further argued that although agents of national security are subject to a number of structural constraints, agents can give direction to structure by creating a need for their own existence, namely by identifying a security threat and then offering their services as a provider of security. Following these initial investigations, the analysis turned to explain the rise of US environmental security. It was argued that this cannot be understood without the context of the end of the Cold War and without recognising the status of environmental issues during that time. The end of the Cold War was the more important of the two. Not only did it provide a window of opportunity for those with a long-standing interest in environmental security issues, but the end of the Cold War itself meant that national security institutions were looking for new missions to justify their continuous existence at a time of dwindling budgets. Environmental security was one such mission.

The analysis of the Clinton administration's first national security strategy and early speeches suggests that the referent object of environmental security was none other than the American people. They were considered existentially threatened in the face of environmental degradation, including global environmental change (climate change and ozone depletion), regional environmental issues (cross-border pollution) and also domestic environmental issues. For the purposes of this book it is vital to note that this security equation (referent object and nature of the threat) did not change in the Clinton administrations' rhetoric over the course of the following years. An extract from the second Clinton administration's NSS from 1999, for example, reads: 'Environmental threats such as climate change, stratospheric ozone depletion, introduction of nuisance plant and animal species, overharvesting of fish, forests and other living natural resources, and the transnational movement of hazardous chemicals and waste *directly threaten the health and economic well-being of US citizens*'.[63]

---

[63] 'A National Security Strategy for a New Century', December 1999, at www.dtic.mil/doctrine/jel/other_pubs/nssr99.pdf[6/2004] p. 13 (emphasis added).

This chapter further showed that while the language of environmental security was all-encompassing, much narrower definitions of environmental security were developed by policy-makers in the Congress. The idea that the military could change its image from being essentially a destroyer of the environment to becoming a good steward of the environment, clean up its heavily contaminated bases and in that way ensure military readiness was one such policy. The security equation of 'defence environmental security' is, however, quite different from that laid out in the securitising move. Thus it is not the American people that was considered threatened by environmental degradation, but rather the DOD's ability to carry out national security. While popular with the Democrats at the Armed Services Committee, and in particular with soon to be Vice President Gore, it is vital to note that defence environmental security did not find its way into the Clinton administrations' national security strategies.

In addition to cleanup missions, the military became involved in environmental security in so far as it was suggested that Cold War intelligence equipment could be used to monitor environmental degradation. The latter was bolstered by the arrival of the environmental scarcity thesis in the policy-making world, which is the idea that environmental scarcity coupled with social effects might lead to overt violent conflict. Unlike defence environmental security, both the possibility of environmentally induced violent conflict as well as 'intelligence environmental security' found their way into the various national security strategies supporting the original securitising move.

4 | *The Clinton administrations and environmental security*

Introduction

This chapter moves from the rhetoric of environmental security to what was done in the name of environmental security, or in other words security practice. As such the chapter is informed by two research questions. First, was the case of United States environmental security under the Clinton administrations a securitisation as opposed to a politicisation? And, if yes, second, why did they securitise? With these questions in mind the chapter analyses the nature of all the policy-making programmes and initiatives that followed the incorporation of environmental security into the NSS, or rather (since the first NSS was only released in 1994) all those that followed the inauguration of the first Clinton administration in January 1993, within all of the relevant government agencies. The relevant government agencies were: the Department of Defense (DOD), the Department of Energy (DOE), the Environmental Protection Agency (EPA), the Central Intelligence Agency (CIA), the Department of State (DOS) and, on the margins, the United States Agency for International Development (USAID). Out of all of these agencies, it was the US Department of Defense (made up of the four components Army, Navy, Air Force and the Marine Corps) that had the biggest profile with regard to environmental security. With the specially created Office of the Deputy Under Secretary of Defense, Environmental Security (ODUSD-ES), the DOD was the only government agency that had a dedicated office for environmental security. Despite the creation of the ODUSD-ES, however, the majority of the programmes that ran from this office were not newly devised under the environmental security mandate, but rather, most were in existence long before environmental security ever became an issue, with many even laid down in federal environmental law. In other words, the DOD was already legally required to conduct at least some of the programmes grouped together under the label 'environmental security'. This alone shows that it is not at all clear whether or not US environmental security was any different from what happened before, or in other words, whether it was a case of securitisation.

Environmental security domestic programmes and initiatives

Department of Defense

In order to understand the circumstances of the creation of the ODUSD-ES it is imperative to consider the historical context in which it was created. When Secretary of Defense Les Aspin took office in 1993 the DOD's operational structure was the same that it had been during the Cold War, leaving Aspin and his staff with the task of modifying the Department's organisation so that it could pursue post-Cold War missions. The reorganisation produced four Under Secretary Offices. One was the Office of the Under Secretary of Defense for Acquisition and Technology, at the time headed by John Deutch. He was responsible, in turn, for a constellation of several Deputy Under Secretaries, an Assistant Secretary, and office directors. One of these was the Office of the Deputy Under Secretary of Defense, Environmental Security. According to Larry K. Smith, who was Counselor to the Secretary of Defense Aspin and to the Deputy Secretary of Defense William Perry, such an office had become necessary to consolidate several large, expensive and consequential environmental initiatives that had emerged and grown over the course of the past years. The creation of the office would give these related and important programmes visibility as an integral set of work, would provide strong, focused leadership invested with an appropriate rank and talent, and would make the entire effort directly accountable to one of the most powerful Under Secretaries. In an interview for this book, Smith further said the office was necessary to 'bring policy coherence for various existing and growing environmental programmes which had been pro-vided billions of dollars for cleanup programmes but which had been producing paper studies only rather than actual cleanup results'.[1]

---

[1] Author telephone interview with Larry K. Smith, former Counselor to the Secretary of Defense Aspin and to Deputy Secretary of Defense Perry, 21 October 2005.

So, Smith maintains, naming the office 'environmental security' was *not* a result of picking a title to reflect an abstract definition of what environmental security meant. Instead the title was a pragmatic adoption of a term from the 'zeitgeist' that captured the substantive common denominator of important existing programmes.

Once the office was created, Sherri Wasserman Goodman, who had staffed the Senate Armed Services Committee in its work on environmental security, was appointed as Deputy Under Secretary of Defense for Environmental Security. The former Deputy Assistant Secretary of the Air Force (Environment, Safety and Health) Gary D. Vest was appointed her deputy. The office was housed in the Pentagon building in Virginia. The following passage gives an idea of how the office defined its goals early on in the administration:

DOD's environmental security strategy will focus on Cleanup, Compliance, Conservation, and Pollution Prevention: C3P2. DOD will provide clear, definitive policy guidance; design an appropriate method of prioritizing, funding, and tracking budgetary requirements; develop innovative technologies by prioritizing and targeting research funds on serious and widespread problems; and increase our outreach to and partnership with other federal agencies, regional and State regulators as well as local communities and foreign governments. International environmental activities will be part of the work.[2]

With environmental security threats defined as follows:

Environmental security threats are conditions affecting human health, safety, or environment that impair DOD's ability to prepare for or carry out the National Security Strategy or create instabilities that can threaten US National Security. They are of three types: *Global:* warming; ozone depletion; loss of biodiversity; proliferation of weapons of mass destruction; international chemical demilitarization. *Regional:* environmental terrorism, accident, or disaster; regional conflicts caused by scarcity/denial of resources; cross-border or global commons contamination. *National:* risks to public health and the environment from DOD activities; increased restriction of military operations; inefficient DOD resource use; reduced weapons system performance; demilitarization of nuclear, chemical and conventional munitions; and erosion of public trust.

[2] Department of Defense Strategy for Environmental Security (1993) unpublished mission statement, courtesy of Deputy Under Secretary of Defense for Environmental Security (1993–2000) Sherri W. Goodman.

Although the ODUSD-ES was created in the spring of 1993 it was not until the spring of 1996 that the office finally summarised its environmental security goals and programmes in an official directive – the DOD 4715.1. This Environmental Security Directive listed a total of eight environmental issue areas, namely: cleanup of military bases; compliance with federal, state and international environmental law; pollution prevention; conservation; planning; technology; international military-to-military cooperation; and education and training.[3] Out of these eight issue areas, cleanup of military bases was easily the best-funded issue, with annually some $2 billion, out of a $5 billion overall environmental security budget, dedicated to this programme alone. The so-called Defense Environmental Restoration Program (DERP) focused on the cleanup of active installations, on so-called formerly used defense sites (FUDS) as well as on installations undergoing base realignment and closure (BRAC). The number of contaminated sites at the time was vast; by the end of 1993, the DOD identified some 21,000 contaminated sites on active and formerly used military bases around the country.[4]

What was left of the overall environmental security budget was committed to the Environmental Quality Program, which encompassed the remaining issue areas, but primarily funded compliance, pollution prevention and conservation. See Table 1 for details.

As already mentioned, the majority of the programmes funded by the ODUSD-ES were *not* new programmes. Thus, cleanup of military bases – under the Clinton administration called DERP – was created in 1975, then, however, under the label Installation Restoration Program (IRP). In addition, since 1980 the DOD has been legally required to provide for contaminated property as a result of past activity. This requirement is fixed in the Comprehensive Environmental Response, Compensation, and Liability Act (CERCLA), also known as the superfund.[5] The CERCLA often applies in conjunction with a second law, the Resource Conservation and Recovery Act

[3] Department of Defense Directive DOD 4715.1 Environmental security, 24 February 1996, at http://biotech.law.lsu.edu/blaw/dodd/corres/pdf2/d47151p.pdf [4/2004].
[4] Stephen Dycus, *National Defense and the Environment* (Hanover: University Press of New England, 1996), p. 95.
[5] CERCLA was amended by the Superfund Amendments and Reauthorization Act (SARA) on 17 October 1986.

Table 1 *Environmental quality budget 1990–2000 (in US dollars)*[6]

Fiscal year	Total budget	Compliance	Pollution prevention	Conservation
1990	-----	790 million	-----	-----
1991	-----	1.108 billion	-----	10 million
1992	-----	1.930 billion	-----	25 million
1993	-----	2.118 billion	274 million	133 million
1994	2.8 billion	1.977 billion	338 million	99 million
1995	2.6 billion	+ 2 billion	284 million	152 million
1996	2.6 billion	2.2 billion	250 million	105 million
1997	2.27 billion	1.90 billion	244 million	108 million
1998	2.3 billion	1.90 billion	256 million	136 million
1999	2.1 billion	1.70 billion	238 million	136 million
2000	2.1 billion	1.66 billion	280 million	165 million

(RCRA), which requires 'corrective actions' to clean up and control hazardous materials already released into the environment.[7] The EPA, since 1992 authorised through the Federal Faculties Compliance Act, together with individual state pollution control agencies, executed these laws, and in cases of non-compliance the DOD and its components were fined. In 1998, for example, the DOD and its components received a total fine of nearly $3 million, with the Army alone accounting for two-thirds of the overall fine.

[6] Data for FY 1994–2000 is taken from the Office of the Deputy Under Secretary of Defense (Environmental Security), 'Defense Environmental Quality (EQ) Programs Annual Reports to Congress', available at www.denix.osd.mil/portal/page/portal/denix/environment/ARC. Data for FY 1990–1993 is taken from the *Report to the Defense Science Board Task Force on Environmental Security* (Washington DC: Department of Defense, 1995). Blank boxes indicate that no data was available. The reason for this becomes clear when considering the following reply to a Freedom of Information Act request (made by the author on 22 January 2005, answered 29 March 2005), the Office of the Deputy Under Secretary for Installations and Environment (I&E) stating that there was 'no formal budgeting or obligation tracking of environmental programmes prior to 1994. Therefore, no responsive documents were located in this search. The office has added that prior to 1994, the financial reporting for these programmes was under the auspices of the various States where the programs were performed.' Without this data the total overall budget obviously cannot be determined.
[7] Dycus, *National Defense and the Environment*, p. 82.

The largest beneficiary of the DOD's environmental quality budget was environmental compliance. From 1994 onwards the programme received an average of almost $2 billion per year, that is, almost the entire second half of the overall environmental security budget. The policy behind environmental compliance was built upon two main pillars. The first pillar aimed to 'ensure that environmental programs achieve, maintain, and monitor compliance with all applicable EOs [Executive Orders] and Federal, State, inter-state, regional, and local statutory and regulatory requirements, both substantive and procedural'.[8] This included environmental laws such as the Clean Air Act, Clean Water Act, Safe Drinking Water Act and the Resource Conservation and Recovery Act. The second pillar aimed to 'reduce compliance costs and simplify requirements to the extent possible'.[9] This was done through pollution prevention, budgeting and environmental planning, whereby the latter was defined as 'the process of identifying and considering environmental factors that impact on, or are impacted by, planned DOD activities and operations'.[10] In short, several of the environmental security issues that fell into the category of 'compliance' were aimed at the prevention of future legislative environmental liabilities. Pollution prevention (the second largest beneficiary of the environmental quality budget), for example, aimed to reduce 'the cost of environmental compliance and helps prevent future liability for environmental cleanup'.[11] Compliance policies applied to the entire military service and its components and to 'all DOD operations, activities, and installations in the United States, its territories, trusts, and possessions, including Government-owned and contractor-operated facilities'.[12]

---

[8] Department of Defense Instruction Number 4715.6, Environmental Compliance, 24 April 1996, p. 2 at www.dtic.mil/cgibin/GetTRDoc? AD=ADA316514&Location=U2&doc=GetTRDoc.pdf [1/2004].
[9] *Ibid.* p. 3.
[10] Department of Defense Instruction Number 4715.9 Environmental Planning and Analysis, 3 May 1996, p. 2 at www.dtic.mil/whs/directives/corres/pdf/ 471509p.pdf [2/2005].
[11] Office of the Deputy Under Secretary of Defense, Environmental Security, '1994 Defense Environmental Quality (EQ) Program Annual Report to Congress' (Washington DC: Department of Defense, 1994) available at www.denix.osd. mil/portal/page/portal/denix/environment/ARC.
[12] *Ibid.*

Comparisons of indicators that measure environmental compliance since the creation of the programme deliver promising results. For instance, compliance enforcement action, the process whereby installations are given the chance to amend possible violations of environmental legislation, instead of being given a fine straight away, declined from 1,200 cases in 1993 to 300 cases in 2000, a decrease of 75 per cent.[13] The level of inspections that led to such fines declined steadily from 38 per cent in 1994, to just 15 per cent in 2001.[14] In addition, the DOD did make some improvements in preventative measures, most significantly in 'pollution prevention'. Hazardous waste production, for example, declined from 1992 with a total 480,000 pounds to 146,527 pounds in 2001, equating to a 69 per cent decline.[15] Similarly, the 'Solid Waste Diversion' rate showed improvements, as in 2001 with 46 per cent, 16 per cent more waste than had been recycled in 1998.[16]

Conservation, as Table 1 emphasises, was the third largest recipient of the Environmental Quality budget. Conservation on military land was, and still is, like cleanup, a legal requirement. The duty to conduct conservation efforts has been fixed with the Sikes Act already in 1960, which requires 'each DOD installation to develop a plan to manage and maintain wildlife, fish, and game conservation and rehabilitation'.[17] Although the Sikes Act gives authority over form and shape of conservation programmes to the Secretary of Defense, the latter is tied to rules and regulations such as the Endangered Species Act (ESA) and the Integrated Natural Resources Management Plan (INRMP) in all conservation efforts. The ESA of 1973 explicitly outlines the obligation of all federal institutions for the conservation of threatened and endangered plants and animals and the habitats in which they are found. The INRMP, the 1997

---

[13] Office of Deputy Under Secretary of Defense, Installations and Environment, '2001 Defense Environmental Quality (EQ) Program Annual Report to Congress' (Washington DC: Department of Defense, 2001), p. 64 available at www.denix.osd.mil/portal/page/portal/denix/environment/ARC.

[14] *Ibid.* p. 64.

[15] Office of the Deputy Under Secretary of Defense, Installations and Environment, 'Pollution Prevention in Progress Review' (Washington DC: Department of Defense, 2002), p. 1.

[16] *Ibid.* p. 2.

[17] Office of the Deputy Under Secretary of Defense, Installations and Environment, '2002 Defense Environmental Quality (EQ) Program Annual Report to Congress' (Washington DC: Department of Defense, 2002), p. 57 available at www.denix.osd.mil/portal/page/portal/ denix/environment/ARC.

amendment of the original Sikes Act, requires that consideration of environmental planning (including resource conservation) is made in all military activity. The DOD Directive 4700.4 on the Natural Resources Management Program released on 24 January 1998 usefully highlights the compatibility between environmental conservation and military missions: 'Natural resources under control of the Department of Defense shall be managed to support the military mission, while practicing the principles of multiple use and sustained yield, using scientific methods and an interdisciplinary approach. The conservation of natural resources and the military mission need not and shall not be mutually exclusive'.[18]

One way in which compliance, pollution prevention and conservation were strengthened further was by educating and training military personnel in environmental security issues. According to the ODUSD-ES's education and training mission statement, the programme was designed so that 'each [military] person can meet his or her environmental responsibilities'.[19] To achieve this goal all military academies and training centres offered compulsory courses for all military personnel in environmental planning. In addition, the DOD advocated an environmental security career development programme, which focused on 'the recruitment, retention and advancement of environmental security professionals'.[20] The DOD invested on average $20 million into the education and training programme annually.

Another way in which the components of the environmental quality programme were supported was the use of technology to meet environmental challenges. Like all of the elements of the environmental quality programme mentioned so far, the DOD's environmental technology programme was designed to cost-effectively meet environmental laws, executives and all other legal requirements, and to protect the department from environmental liabilities.[21] Funding for the

[18] Department of Defense Directive 4700.4, Natural Resources Management Program, 24 January 1998, p. 2 at http://biotech.law.lsu.edu/blaw/dodd/corres/pdf/d47004_012489/d47004p. pdf [1/2005].

[19] Office of the Deputy Under Secretary of Defense, Environmental Security, 'Mission Statement Environmental Security Education' (Washington DC: Department of Defense, 1994).

[20] *Ibid.*

[21] Office of the Deputy Under Secretary of Defense, Environmental Security, '1996 Defense Environmental Quality (EQ) Program Annual Report to Congress'

environmental technology programme made up on average about one per cent of the overall Environmental Quality budget and was provided for by the other elements of that programme. The pollution prevention initiative, for instance, provided some of the money to advance research into new technologies that facilitated pollution prevention. Most of the environmental technology fund was dedicated to two environmental technology programmes: (1) the Strategic Environmental Research and Development Program (SERDP) for science and technology, and (2) the Environmental Security Technology Certification Program (ESTCP) for demonstration and validation. These two environmental technology programmes were far too complex to be described here in any detail, but with regard to defence environmental security they were concerned with improving mission readiness through environmental research.

This ends the analysis of the domestic side of the DOD's environmental security programme. Given the successes the DOD clearly had in all of the designated areas, and in particular with regard to cleanup, compliance and pollution prevention, the programme must be viewed in a positive light. Nonetheless, for the purposes of this book, the question remains: what was different about these programmes in comparison to the ones in place before it was called environmental security? Or, in other words: what precisely constitutes a change of behaviour on the part of the DOD? The answer to this question consists of several different elements, all of which are important. The new programme constituted a change of behaviour on the part of the DOD, first, because of the creation of an actual Office of the Deputy Under Secretary of Defense for Environmental Security complete with a mandate – the DOD Directive 4715.1 – for environmental security. Second, because of the difference in funding; thus the creation of the ODUSD-ES went hand in hand with the creation of a stable $5 billion annual budget for the office initially set out for five years – some $25 billion. The DERP alone, for example, grew from just $601 million in 1990 to $2.3 billion in 1994.[22] Third, not only did the money increase significantly (see Table 1), but also from 1993 onwards financial reporting of all

(Washington DC: Department of Defense, 1996), ch. 6, available at www.denix. osd.mil/portal/page/portal/denix/environment/ARC.

[22] Office of the Deputy Under Secretary of Defense for Environmental Security, '1994 Defense Environmental Restoration Programme (DERP) Annual Report

existing defence environmental programmes was taken away from the individual states where programmes were performed and became centrally controlled by the DOD, ensuring that something was being done. Fourth, new initiatives like, for example, the 1994 Pollution Prevention Strategy were developed. Fifth, the individuals within the DOD, Sherri W. Goodman and her deputy Gary Vest, took a very strong interest in the new programme of the DOD.[23] In an interview for this book, Goodman explained that 'the office of environmental security had a much broader mandate than anything in existence before, with a lot of people, ready to make a difference'.[24] Moreover, Vest, had not only a strong interest but years of experience with regard to environmental stewardship of the military, as he had been heavily involved in the subject matter in both NATO and the Air Force. Not only did the ODUSD-ES leadership have a strong interest in environmental security, but, as already shown in the previous chapter, political leaders such as the Vice President actually helped to take the programme off the ground, time and again highlighting the link between environment and security. To summarise, although at first glance somewhat similar to what went on beforehand, the domestic environmental security programme of the DOD differs significantly with regard to budget, leadership and commitment – suggesting a securitisation as opposed to a mere politicisation of the environment.

Environmental security international programmes and initiatives

Department of Defense, Environmental Protection Agency and Department of Energy

The eighth and final environmental issue featured within the DOD's Environmental Security Directive was international military-to-military

---

to Congress' (Washington DC: Department of Defense), available at www.denix.osd.mil/portal/page/portal/denix/environment/ARC.

[23] Regarding the commitment of Goodman *et al.* see also Robert F. Durant, *The Greening of the US Military: Environmental Policy, National Security, and Organizational Change* (Washington DC: Georgetown University Press, 2007), pp. 8, 52–73, 245.

[24] Author interview with Under Secretary of Defense for Environmental Security (1993–2000) Sherri D. Goodman, 13 September 2005, Alexandria VA.

cooperation, henceforth also referred to as international environmental security. This programme is important for two different reasons: first, it extended the narrow mandate of the ODUSD–ES significantly; and, second, it brought a number of other government agencies into the environmental security story.

Although the notion of international environmental security was part of the DOD's environmental security initiative from the very beginning, it was not until the commencement of Secretary Perry's strategy of 'preventative defence' that the international remit of environmental security gained importance.[25] The strategy was based on the belief 'that the best security policy is one which prevents conflict'.[26] In the 1990s this strategy focused in large part on the promotion of democracy worldwide; on par with the teachings of the democratic peace thesis, democratisation was believed to reduce the chances of interstate conflict. In practice, throughout the 1990s, 'preventative defence' required the DOD to engage in missions such as conflict prevention, arms reduction, prevention of proliferation, cooperation with militaries in aspiring democracies, and peacekeeping. For environmental security, according to Goodman, preventative defence meant two things:

One challenge is to understand where and under what circumstances environmental degradation and scarcity may contribute to instability and conflict, and to address those conditions early enough to make a difference. The second challenge is to determine where environmental cooperation can contribute significantly to building trust and understanding. These two elements together constitute the environmental security pillar of 'preventative defense'.[27]

The first of these two points is of course an acknowledgement of the linkage between violent conflict and environmental scarcity. As the previous chapter has shown, much rhetorical attention was given to this linkage, with policy-makers from various government agencies echoing the environmental scarcity thesis. In practical policy-making

---

[25] Secretary of Defense Les Aspin was replaced by Deputy Secretary of Defense William J. Perry after the former's early resignation on 3 February 1994.

[26] William J. Perry, 'Good Stewards at Home, Good Stewards Abroad', Remarks to John F. Kennedy School of Government, Harvard University, 13 May 1996 at www.loyola.edu/dept/politics/intel/goodman.html [1/2005].

[27] Sherri W. Goodman, 'The Environment and National Security', speech at the National Defense University, 8 August 1996 at www.loyola.edu/dept/politics/intel/goodman.html [1/2005].

terms, however, the efforts were, especially compared to the strong rhetorical commitment, rather small. Thus, the DOD for its part mainly focused on the organisation of conferences and working groups, where the possibility of environmentally induced conflict was being discussed. The same is true for the CIA's Directorate for Central Intelligence Environment Center, which analysed research on the environment and conflict, organised workshops on the subject and arranged a gaming exercise at the US Army War College in anticipation of the 1997 climate change negotiations in Kyoto.[28] Thomas Homer-Dixon, along with numerous other academics, got to work on the 'State Failure Task Force', a CIA-sponsored multi-actor research group, initiated by Vice President Gore and put together by William (Bill) Wise, the Deputy National Security Advisor to the Vice President. It must be noted that although Homer-Dixon was a member of the State Failure Task Force, this group was *not* put together because of the link between environment and security, but because of the many failed states during the early 1990s and the wish of the Clinton administrations to better understand the reasons for this failure.[29] Upon reading the early reports produced by the group, however, it becomes clear that environmental degradation was at least considered one variable in state failure.[30]

Unlike the first connection, the second connection between preventative defence and environmental security has so far hardly been mentioned. This is particularly interesting, as it was this programme that significantly broadened the mandate of the ODUSD-ES, making this programme one of the most interesting initiatives in the overall environmental security effort of the Clinton administrations. In what follows, this programme and its outcome will be examined in detail.

The end of the Cold War left not only the US military but also the now numerous states of its former enemy the Soviet Union with a

[28] Geoffrey D. Dabelko, 'Tactical Victories and Strategic Losses: The Evolution of Environmental Security', unpublished doctoral thesis, Faculty of the Graduate School of the University of Maryland (2003), p. 75.

[29] Interview with the National Security Advisor to Vice President Al Gore (1993–2001) Leon Fuerth, 12 September 2005, Washington DC.

[30] Homer-Dixon himself was not impressed with the work of the State Failure Task Force. He participated in no more than two meetings, and in 1996 was invited to publicly critique the results of the group's findings at a Woodrow Wilson Center meeting.

detrimental environmental legacy. If anything, due to the lack of concrete plans as to how to handle decommissioned nuclear submarines, coupled with a legacy of uncontrolled dumping of nuclear waste from such submarines, these states faced an even grimmer environmental future than the US. The American Director of the Arctic Military Environmental Cooperation programme (AMEC) Dieter K. Rudolph describes the environmental practices of the Soviet Union during the Cold War as follows:

The Soviet Union dumped radioactive waste from nuclear submarine operations in the oceans since 1960. Although other nations have also dumped radioactive waste at sea, the Soviet Union dumped more than twice as much radioactivity as other countries. The Bellona Foundation estimates that through 1994 the Russian Northern Fleet sank seventeen ships and three barges containing radioactive waste in the Barents and Kara Seas. Russian sources estimate that the total amount of waste dumped was between 11,000 and 17,000 containers of solid waste and 165,000 cubic meters of liquid waste. In 1993 Russia reported that the Soviet Union sank sixteen nuclear reactors in the Kara Sea and three reactors in the Sea of Japan. Six of the reactors dumped in the Kara Sea were so badly damaged and radioactive that they were dumped with the nuclear fuel.[31]

Given the nature of the environmental legacy of the Cold War many western states, including the US but also particularly Norway, adjacent to Russia, began to fear a new threat to security, this time, stemming from the environmental contamination of nuclear waste, as opposed to that posed by hot nuclear warheads. Furthermore, the US in particular feared that the unsecured nuclear weapons would fall into the wrong hands, and foster the proliferation of weapons of mass destruction. The threat of nuclear waste dumping became particularly pervasive in 1993 when a white paper written by the Russian Yablokov Committee, which was in charge of analysing the environmental situation in Russia at the time, openly accused the Russian government of the illegal dumping of nuclear waste in the Murmansk area. Afraid of the possibly devastating consequences for their own country, Norway sought help from the American government, with the Norwegian Defence Minister Jørgen Hårek Kosmo asking the US

---

[31] Dieter Rudolph, 'Arctic Military Environmental Cooperation AMEC: An Effort of the International Defense Community to protect Coastal and the Marine Environment in the Russian Arctic', unpublished manuscript (2004), p. 2.

Secretary of Defense William Perry for help during a 1995 meeting.[32] Secretary Perry acted immediately and entrusted Goodman and her team with the issue. Work began without delay, even though it was not till 1996 that a formal agreement, the Arctic Military Environmental Cooperation (AMEC), was signed between Russia, the US and Norway.

For the US and Norway the way to do something about this environmental threat was to offer cooperation on environmental issues, particularly on nuclear waste management, to the Russian counterpart. At this point, it is important to remember the historical context in which all of this took place. The Cold War had ended, the United States was the victorious power, and relations with the former foe Russia were sensitive. It is therefore not surprising that international military-to-military cooperation was not only regarded as the solution to the Cold War military's environmental problems, but also as a way to build peace, trust and stability between the former enemies. In other words, the environmental problems in Russia, as bad as they may have been, came with the benefit of offering an entry point for cooperation and mutual trust building between the former enemies.

That the Russian threat as well as the peace-building potential of environmental cooperation was taken seriously becomes evident when considering that, at one point, there were up to three different programmes in existence in the Murmansk area alone. These three were AMEC, environmental initiatives under NATO's Committee on the Challenges of Modern Society (CCMS)[33] and the programmes under the so-called Gore–Chernomyrdin Commission (GCC) established in April 1993, formally called the US–Russian Joint Commission

---

[32] Author interview with Sherri W. Goodman and author telephone interview with American Director of the Arctic Military Environmental Cooperation (AMEC) Dieter Rudolph, 27 October 2005.

[33] In the final report to a pilot study entitled 'Environment and Security in an International Context' the aim of CCMS is described as follows: 'to tackle challenges of modern society, particularly problems affecting the environment of the nations and the quality of life of their peoples, taking advantage of the potential for co-operation offered by the Alliance. [...] Two important concepts characterise the work of the Committee, namely that it shall consider specific problems of the human environment with the deliberate objective of stimulating action by member governments and that its final results should be entirely open and accessible to international organisations or individual countries elsewhere in the world': Gary D. Vest and Kurt M. Lietzmann, *Environment and Security in an International Context* (Washington DC: Department of Defense, 1999), p. 3.

on Economic and Technological Cooperation. However, while in theory there existed clear dividing lines between the different environmental initiatives in the Murmansk area, in practice they were very much blurred. According to former Principal Assistant Deputy Under Secretary of Defense for Environmental Security Gary Vest, 'The American teams in Russia under AMEC, CCMS and GCC were roughly all the same people. All of us would meet periodically and talk about AMEC and GCC, because all of our agencies were involved. One staffer, Wendy Grieder, for example, was the EPA's principal staffer on the MOU, but she was also in charge of NATO's Russian CCMS programme. From a practitioner's standpoint it did not matter under which programme what was done'.[34]

Out of the three programmes AMEC is the most noteworthy. It focused primarily on the safe storage of spent nuclear fuel retrieved from Russia's decommissioned fleet of nuclear submarines. The programme initially evolved out of the Arctic Nuclear Waste Assessment Program (ANWAP). During its first year, 1997, the programme was financed by the Nunn–Lugar Cooperative Threat Reduction Program that oversaw a budget of some 500 million US dollars, all of which was set aside for the destruction of weapons of mass destruction (particularly MIRVs and ICBMs) as outlined under START II. In subsequent years funding was to come directly from Congress. However, the national security objective – the reduction of nuclear warheads, to counter the proliferation of nuclear weapons – remained the same and AMEC's efforts were in direct response to these national security objectives.

Although on the US side the DOD, in the form of the ODUSD-ES, was the main player in AMEC, the DOD had little or no experience in nuclear waste management and lacked the mandate to act as they pleased, which made it necessary for other agencies to be invited to contribute. These agencies were the US Environmental Protection Agency (EPA) and the Department of Energy (DOE). For the EPA the environmental security mission was to become a mission much liked and sought after. This was because the EPA regarded environmental security as a springboard to become a bigger player

---

[34] Author interview with Principal Assistant Deputy Under Secretary of Defense for Environmental Security (1993–2001) Gary D. Vest, 17 September 2005, Arlington VA.

within the US government, and more importantly as the key to end notorious under-funding.[35] The pivotal contribution of the agency to AMEC was the purposeful construction of mobile concrete casks (first with a capacity of eight tons expanding to a capacity of forty tons later), designed to transport spent nuclear fuel from the Murmansk region to the Mayak reprocessing facility in the Urals. This project became the centrepiece of AMEC, catapulting the EPA to being one of the most important actors in US international environmental security efforts.

For the DOE, environmental security was a welcome and new initiative as well. Like the EPA, they saw environmental security as a means to gain funding and maintain their status within the US government.[36] As already noted (Chapter 3), the nuclear management of the DOE had come under criticism in the US itself after the end of the Cold War, leaving the agency constantly in fear of being down-sized or replaced altogether. Environmental security was a chance to remain in the game. The DOE's involvement in AMEC was necessary because of the agency's undisputable expertise in nuclear waste management, and because of its unmatched resources such as the national weapons laboratories, all of which are owned by the DOE. Under AMEC, the agency's biggest tasks lay with the training of Russian workers in nuclear waste management and in how to safely dispose of nuclear waste.

As with the other two agencies, AMEC was very popular within the ODUSD-ES. For Goodman and her team, AMEC constituted the first real opportunity to do something about the international dimension of defence environmental security, in the process extending the office's mandate. Besides providing the majority of the needed resources – the DOD under the second Clinton administration spent a total of

---

[35] Author interview with the Principal Deputy and Deputy Assistant Administrator for International Activities at the EPA (1989–2001) Alan Hecht, 13 September 2005, Washington DC, and author interview with the Assistant Administrator for International Activities at the EPA (1994–2001) William Nitze, 15 September 2005, Washington DC; See also William Nitze, 'A Potential Role for the Environmental Protection Agency and Other Agencies', *Environmental Change and Security Project Report* (Washington DC: The Woodrow Wilson Center, 1996), p. 118.

[36] Author interview with Deputy Assistant Secretary for Energy, Environment and Economic Policy Analysis (1990–1999) Abraham Haspel, 16 September 2005, Washington DC.

$4–$6 million annually on AMEC – the entire initiative was coordinated and led from Goodman's office, making the DOD the most important actor in environmental security domestically, as well as internationally.

Notwithstanding the leadership of the DOD, the contribution of the other institutions was vital and was recognised as such within the DOD. Given the successful working relations between the three agencies – DOD, EPA and DOE – under AMEC, coupled with all of the agencies' desire for international environmental security, AMEC was only the starting point for further international environmental security missions conducted jointly by the three agencies. Missions aside, clearly the most outstanding diplomatic success, and therefore success, at least in terms of recognition, for environmental security, was the creation and quick signing (in July 1996) of a Memorandum of Understanding (MOU) among the EPA, DOE and DOD concerning cooperation with regard to environmental security.

To publicise the MOU, the EPA prepared a brochure entitled *EPA Strategic Plan for Environmental Security*. The brochure drew the attention of the conservative *Washington Times*, which lamented that the EPA and the brochure were more evidence of the Clinton administration's 'wacky environmental policy'.[37] Even though articles such as this one did not actively change anything in policy-making terms, they were a blow to the internationally minded people within the EPA, particularly as their boss, Carol Browner, who herself was much more interested in the state of the domestic environment anyway, was hardly amused by the bad press.[38]

Besides the conservative press, the MOU and particularly the subsequent 'going it alone' by the DOD was not popular within the Department of State. Initially the MOU was put forward for signing by the DOS as well, but the Assistant Secretary of State for Oceans and International Scientific Affairs Eileen Claussen refused to sign the memorandum. According to Dabelko: 'She [Claussen] had earlier expressed her poor estimation of the environment and conflict work done by Homer-Dixon. Her primary focus was the negotiation of international environmental treaties such as the Framework Convention on Climate Change [...] More parochially, the MOU would grant

---

[37] Author interview with Alan Hecht.
[38] Author interviews with William Nitze and Alan Hecht.

the resource-rich and active Defense officials some of the diplomatic authority that the Department of State guarded jealously'.[39]

The view that the DOD and the DOS closely guard their diplomatic influence is widely shared in Washington and it is a well-known fact that traditionally there is rivalry between the two agencies. The DOS, of course, has another take on the issue and, according to a confidential source from the State Department, the MOU was regarded as 'a subset to its own environmental diplomacy initiatives'. This was why the MOU always had to be accompanied by a written statement issued by Secretary of State Warren Christopher, which praised the efforts of the three agencies as timely, but left no doubt about the special status regarding foreign policy issues claimed by the DOS. In the long run, however, these interagency quarrels meant that the MOU suffered as a result. To cite Dabelko once more: 'Persistent middle level bureaucratic opposition in the Department of State's OES undercut the effectiveness of the MOU and all efforts by the MOU parties to create more permanent bureaucratic environmental security offices within the US government'.[40]

In the short term, however, the MOU could account for some successes in international environmental security efforts. To understand the significance of the MOU properly, it is useful to consider what conception of environmental security it was working with. Principal Deputy and Deputy Assistant Administrator for International Activities at the EPA (1989–2001) Alan Hecht provided the following definition:

One is the National Security Strategy. A quote from the report states that 'even when making the most generous allowance for advances in science and technology, one cannot help but feel that population growth and environmental pressures will lead to immense social unrest and make the world substantively more vulnerable to serious international pressure'. We are now trying to specifically address those 'environmental pressures'. The legacy of the Cold War was another stimulus in our case. The legacy meant that the management of radioactive chemical and biological facilities, the transition of what were formally military to civilian facilities, and the various other problems associated with the democratisation processes all contributed to environmental security issues. We could see that these issues were only going to get more serious because implementation of the SALT agreements meant

---

[39] Dabelko, 'Tactical Victories and Strategic Losses', p. 73.
[40] *Ibid.*

the decommissioning of greater numbers of nuclear submarines and the generation of greater quantities of liquid and solid waste.[41]

Considering that this definition – especially the second part of it – came with a cohesive policy-making plan, it is not at all surprising that for a number of former Clinton officials the MOU, at least in its theoretical ambit, *was* environmental security. In practice, and as already mentioned, the MOU was destined to be less successful than its theoretical framework promised. Nevertheless, cooperation between the EPA, the DOE and the DOD under the MOU can account for some small successful international environmental security initiatives. Thus, the three agencies cooperated to promote a diverse number of environmental goals: for example, environmental diplomacy, environmental training, environmental cleanup, and the prevention of environmental crime in environmental hot spot regions, including among other regions the Baltic Sea Region, the Middle East, Panama and the Arctic.[42]

Stimulated by this step into the international dimension of environmental security, but at the same time disillusioned with the mediocre success of the MOU, Gary D. Vest in his role as Principal Assistant Deputy Under Secretary of Defense for Environmental Security, who was according to his contract in charge of the international side of defence environmental security, started his 'own' international initiatives. During the years from 1997 to 1999 Vest, with the approval of his colleague and superior Goodman and her superiors, travelled the globe building relations between the US military and foreign militaries. Although the environment, or rather the benefit of the military for the environment, or as Vest put it 'making the world's militaries environmentally sensitive',[43] was the initial motive of this military-to-military cooperation, Vest claims to have soon realised that this was merely a tool for grander motives such as peace, trust building and

[41] Hecht cited in Abraham Haspel, Alan Hecht and Gary Vest, 'The DoD–DoE–EPA Environmental Security Plan', *Environmental Change and Security Project Report* (Washington DC: The Woodrow Wilson Center, 1997), pp. 163–4.
[42] US Environmental Protection Agency, *Environmental Security: Strengthening National Security through Environmental Protection* (Washington DC: Environmental Protection Agency, 1999).
[43] Haspel, Hecht and Vest, 'The DoD–DoE–EPA Environmental Security Plan', p. 163.

democratisation. In a sense, therefore, Vest or rather the DOD were doing their own version of foreign policy, often sidestepping more traditional diplomatic channels. Vest, however, with his somewhat unusual code of conduct was successful and, during the administration's remaining three years, he built military-to-military relations with an array of different countries, including the Philippines, the Arabian Gulf States, South Africa, the Czech Republic, the European Union, Argentina, Chile, Australia, Canada, the Baltic countries and the states of the former Yugoslav Republic.[44] These military-to-military programmes took the following form: 'Delegation exchanges; joint analysis of environmental data; information sharing; bilateral or multilateral development of ESOH products, such as handbooks, which are generic in nature and can be utilised in promoting ESOH concepts in militaries worldwide; and hosting or attending conferences that address military ESOH issues in a regional or multilateral context'.[45]

Simultaneously, US Central Command (CENTCOM) under the leadership of General Anthony Zinni also took an interest in environmental security. Although CENTCOM had no dedicated environmental security budget, Zinni and his staff recognised environmental problems as potential future security problems. CENTCOM's 'area of responsibility' (AOR) includes the Middle East, Central Asia and East Africa. In all of these regions, water access, quality and control were considered the leading environmental security issue, which is to say the issue most likely to lead to violent conflict. Like Vest, however, Zinni strongly believed that environmental problems could be utilised to foster cooperation between countries, and in the years 1997–2000 CENTCOM worked closely with the ODUSD-ES and the EPA on the organisation of a number of conferences and workshops in the AOR, with the purpose of educating foreign militaries in environmental awareness.[46]

Department of State

The connection between foreign policy and environmental security makes it necessary to move on to the traditional foreign policy

[44] Office of the Deputy Under Secretary of Defense, Environmental Security, '2000 Defense Environmental Quality (EQ) Program Annual Report to Congress' (Washington DC: Department of Defense, 2000), pp. 49ff., available at www.denix.osd.mil/portal/page/portal/denix/environment/ARC.
[45] *Ibid.* p. 48.
[46] Author telephone interview with General Anthony Zinni, 18 June 2008.

institution within the US government, the State Department. As was the case for the DOD with the creation of the ODUSD-ES, the advent of the Clinton–Gore administration in 1993 meant an institutional change in environmental policy-making terms for the State Department. The new administration created the Office of the Under Secretary of State for Global Affairs (OUSGA), headed by Gore's longtime friend the committed environmentalist Senator Timothy E. Wirth. Wirth's office was entrusted with the oversight of four global issue areas previously subordinated to other agencies: 'Political Affairs Management [human rights and democracy], Arms Control and International Security, Economics, Business & Agriculture, and Global Affairs'.[47] Wirth was assisted by a group of four assistant secretaries working within different areas of concern, namely within the Bureau of Democracy, Human Rights and Labor; the Bureau of Populations, Refugees and Migration; the Bureau of Oceans and International Environmental and Scientific Affairs; and the Bureau of International Narcotics Matters, respectively. Given the size of the OUSGA and the pooling of resources to come to terms with these global issues, the Clinton–Gore administration sent out a strong signal and they appeared genuine in their interest in and commitment to the new global issues. Whether or not this interest/commitment translated into the securitisation of the environment on the part of the DOS, however, is another question. In trying to answer this question it is important to remember that a securitisation does not require the usage of the actual word 'security'. As Wæver explains:

In practice it is not necessary that the *word* security is spoken. There can be occasions where the word is used without this particular logic at play, and situations where it is metaphorically at play without being pronounced. We are dealing with a specific logic which usually appears under the name security, and this logic constitutes the core meaning of the concept security, a meaning which has been found through the study of actual discourse with the use of the *word* security, but in the further investigation, it is the specificity of the rhetorical structure which is the criterion – not the occurrence of a particular word.[48]

[47] Thomas Lippman, 'With Tim Wirth in Position, The Old Lines Lose Weight', *Washington Post,* 30 June 1994, p. 1.
[48] Ole Wæver, *Concepts of Security* (Copenhagen: Institute of Political Science, University of Copenhagen, 1997), p. 49.

According to Wirth's own admission he believes and believed all along that the environment is a security issue, understood in accordance with the definition found in the days of the Clinton national security strategies.[49] The following quote taken from one of Wirth's speeches from that time highlights his conviction well:

Our deficit spending of environmental capital has a direct, measurable impact on human security. Simply put, the life support systems of the entire globe are being compromised at a rapid rate – illustrating our interdependence with nature and changing our relationship to the planet. Our security as Americans is inextricably linked to these trends. The security of our nation and our world hinges upon whether we can strike a sustainable, equitable balance between human numbers and the planet's capacity to support life.[50]

With Wirth as head of the OUSGA committed to such a broad understanding of environmental security, the office followed suit and the first few years of Wirth's period in office were spent promoting one global issue after the other, always in line with what was on top of the global agenda.[51] For example, in the run-up to the 1994 Populations Conference in Cairo, Egypt, Wirth and his staff gave numerous speeches pointing to the potential dangers of overpopulation and possible solutions to the problem. All of these closely resembled a broad understanding of environmental security. For example:

The Program of Action [developed at Cairo] calls for a comprehensive approach that embraces the provision of family planning and reproductive health services, the education and empowerment of women, improved maternal and child health, and the mobilisation of institutional and financial resources. All these initiatives influence population growth. The program also acknowledges that both rapid population growth and wasteful resource consumption play major roles in environmental degradation.[52]

Despite Wirth's heavy campaigning on various environmental issues, and his own conviction of a broad notion of environmental security,

---

[49] Author interview with Under Secretary of State for Global Affairs (1993–1997) Timothy E. Wirth, 21 September 2005, Washington DC.

[50] Timothy E. Wirth, *Sustainable Development and National Security*, (Washington DC: Bureau of Public Affairs US Department of State, 1994), p. 4.

[51] It should also be noted here that Jessica Tuchman Mathews, author of the influential article 'Redefining Security', was employed as Wirth's deputy.

[52] Timothy E. Wirth, *Environmental Challenges Confront the Post-Cold War World* (Washington DC: Office of the Under Secretary of State for Global Affairs, 1995), p. 2.

however, neither the DOS nor the OUSGA ever propagated an actual policy of environmental security. And it was not until 1997 that they settled for the official term of 'environmental diplomacy' to describe what they were actually doing and finally outlined a policy-making plan. Given the aim of this chapter, it must be asked, what was environmental diplomacy? And how did it differ from environmental security?

The official 1997 statement called *Environmental Diplomacy: The Environment and US Foreign Policy* described the purposes of environmental diplomacy as follows:

The State Department now operates on the premise that countries sharing common resources share a common future and that neighboring nations are downstream and upwind, not just north and south or east and west, of each other. Threats to a shared forest, a common river, or a seamless coastline are forcing countries to expand their existing bilateral relationships to include environmental issues, and to create new regional frameworks to confront and combat shared environmental challenges.[53]

Environmental diplomacy was therefore the idea that environmental concerns and threats had become part of the diplomatic relations with other countries. According to the State Department's 1997 report, there were three reasons why the integration of environmental issues into foreign policy was necessary: (1) to help stabilise a region where pollution or the scarcity of resources contributes to political tensions; (2) to enable the nations of one region to work cooperatively to develop initiatives to attack regional environmental problems; (3) to strengthen our relationship with allies by working together on internal environmental problems.[54]

When reading this one cannot but notice the close resemblance between some of what was proposed here and environmental security. Thus, the possibility of environmentally induced conflict was the reason why help was needed in the first instance, whilst environmental trust and peace building is describing the desired form of help. Reason number two – 'to enable', in the widest sense – also falls within the realm of environmental security. However, 'enabling'

[53] US Department of State, *Environmental Diplomacy: The Environment and US Foreign Policy* (Washington DC: Department of State, 1997) at www.state.gov/www/global/oes/earth.html [7/2005].
[54] *Ibid.*

here is understood more concretely as 'sustainable development', a concept with an agenda of its own that is not subordinate to environmental security. In the light of these findings, then, it can be argued that environmental diplomacy in many ways *was* environmental security, simply under a different name. This said, it is interesting to note that Wirth actually wanted to name things by what they were, and used the term environmental security. According to his own admission, however, he had little luck with this label amongst the overwhelming majority of 'traditionalists' within the National Security Council, who thought it more important to have missiles on three-minute alert.[55]

The DOS pursued two pathways to integrate its environmental diplomacy mission. The first pathway was the establishment of so-called regional environmental hubs, created within twelve US embassies in designated environmental hotspot regions.[56] Work within these hubs aimed to foster regional environmental cooperation through data exchange, transparency, pooling of resources etc. between the different countries and the US. The second pathway was to raise the profile of environmental issues in bilateral and multilateral and international relations. A good example of the rise of the profile of environmental issues in multilateral relations is the recognition of the role of water in the Middle East peace process.

Much has been written in the academic literature on the linkage between water scarcity and violent conflict.[57] Water is seen as both a strategic goal for warfare and a strategic weapon in the conduct of war, with many arguing that so-called 'water wars' will come to dominate the wars of this millennium. In recent years, much of this literature has become subject to rigorous empirical analysis and much

---

[55] Author interview with Tim Wirth.

[56] Namely: Addis Ababa, Ethiopia; Amman, Jordan; Bangkok, Thailand; Brasilia, Brazil; Budapest, Hungary; Copenhagen, Denmark; Gaborone, Botswana; Kathmandu, Nepal; Libreville, Gabon; San José, Costa Rica; Suva, Fiji; Tashkent, Uzbekistan.

[57] See, for example, Norman Myers, *Ultimate Security: The Environmental Basis of Political Stability* (New York: Norton, 1993); Peter H. Gleick, *The World's Water: The Biennial Report on the World's Fresh Water Resources* (Washington DC: Island Press, 1989); Aron Wolf, '"Water Wars" and Water Reality: Conflict and Cooperation along International Waterways', in Steve C. Lonergan (ed.), *Environmental Change, Adaptation, and Security* (Dordrecht: Kluwer Academic Publishers, 1999), pp. 251–65.

criticism. Steve C. Lonergan, for example, has argued that it is difficult to single out water as the single variable in any conflict, and that much of the evidence is simply anecdotal.[58] A popular example used to 'prove' the existence of water wars is the 1967 conflict between Jordan and Israel. Norman Myers, for example, has argued that 'Israel started the war in part because the Arabs were planning to divert the waters of the Jordan River system',[59] whereas Lonergan disputes such claims and argues that 'there is little evidence – other than hearsay – that water played a major role in the 1967 war'.[60] In spite of the nature of this academic dispute, it remains a fact that, since the 1967 war, water issues have become politicised, if not securitised, in the Middle East. This politicisation, combined with the environmental diplomacy thinking of the DOS, led to the incorporation of water into the 1991 Madrid peace process negotiations, with the 'Working Group on Water Resources' and notably the 'Working Group on the Environment', alongside more traditional working groups such as the 'Working Group on Arms Control and Regional Security', accounting for two of the five multilateral working groups. The nature of the multilateral cooperation was such that the water group was cosponsored by the US and Russia, with the former acting as gavel holder. Co-organised by the EU and Japan, it involved the following regional actors: Algeria, Bahrain, Egypt, Israel, Jordan, Kuwait, Mauritania, Morocco, Oman, Palestinians, Qatar, Saudi Arabia, Tunisia, United Arab Emirates, Yemen, as well as two dozen delegations from other countries, plus the World Bank. The environment working group was of a similar scale.

Apart from raising the environmental profile of such regional issues, the DOS also sought to raise the profile of global environmental issues. Thus, as already mentioned, Wirth did much to promote population issues in the run-up to the 1994 Cairo conference, but he also promoted sustainable development and global warming; the latter, particularly in the run-up to the 1999 Kyoto negotiations was high on the DOS's agenda. Global warming was particularly close to the interest of the Vice President and therefore

[58] Steve C. Lonergan, 'Water and Conflict: Rhetoric and Reality', in Paul F. Diehl and Nils Petter Gleditsch (eds.), *Environmental Conflict* (Oxford: Westview Press, 2001), p. 119.
[59] Myers, *Ultimate Security*, cited *ibid.* p. 119.
[60] *Ibid.* p. 119.

very much part of the administration's global issues agenda from the beginning.

Given all that has been said here about the nature of environmental diplomacy and its implementation, it appears that the DOS's efforts can only be viewed in a positive light. Such a conclusion, however, might be too rushed. No matter how good environmental diplomacy looked in theory, it did not leave everyone involved satisfied, and in practice environmental diplomacy under the Clinton administrations was subject to a turf war between Wirth and Clinton's first Secretary of State Warren Christopher. Because the dynamics of this turf war throw a slightly different light onto the commitment of the DOS (here excluding Wirth's office) and, potentially, that of the White House to issues of environmental diplomacy, it is worthwhile to outline these dynamics in some detail.

'Officially' the turf war began with an unofficial foreign policy statement speech by Secretary of State Christopher delivered at Harvard University on 22 January 1995.[61] Christopher's speech caused outrage on the part of Wirth as global issues took up no more than a paragraph in a speech that was several pages long. Frustrated with the Department's official policy line, Wirth launched a counter-attack in the public press, in which he accused the State Department of clinging 'to "the old culture [where] real men do politics and arms control" not the environment and human rights'.[62] The DOS defended its line of foreign policy-making by stating that 'nothing on Wirth's agenda is on the forefront of Christopher's concerns for 1995, a crucial year for nuclear nonproliferation, Middle East peace negotiations and planning the future of NATO'.[63] A year later, however, in which he visited several

---

[61] The term 'officially' is used here, because this is the event commonly regarded as the starting point between the falling out between Wirth and his superiors. This said, however, it should be noted that Wirth, or rather his office, did not get off to an excellent start in the first place. Not only did it take Congress more than a year to create a new, $120,000-a-year undersecretary position in the State Department (Terry Atlas, 'Tim Wirth Takes on the World of Problems', *Chicago Tribune*, 7 September 1994), but also his deputy Jessica Tuchman Mathews resigned after just one year in the job, frustrated with the slow progress of the Office for Global Affairs.

[62] Wirth cited in Thomas Lippman, 'Tim Wirth versus State', *Washington Post*, 20 April 1995.

[63] *Ibid.*

countries with very high levels of environmental degradation, Christopher seemed to have come around to Wirth's way of thinking.

> I kept running into political or security problems that had a very large environmental content [...] Haiti stuck out in my mind, with the overpopulation and the deforestation of the country. And in Eastern Europe, those new democracies are struggling with a legacy of environmental abuse and are never going to recover [...] I'd have to say, I was considerably influenced by my meetings with Vice President Gore. I got the feeling that although Wirth's department was operating well, it was not having a pervasive enough influence on the rest of our diplomacy.[64]

In a speech on 9 April 1996 at Stanford University, California, Christopher then made public amends to Wirth, calling for 'a special emphasis on environmental protection, arguing that global peace and national security are increasingly dependent on the health of the world's natural resources'.[65] The speech resulted in much praise of Christopher in the press, and it went in conjunction with the authorisation of Wirth to appoint his favourite, the veteran environmentalist Eileen Claussen, as Assistant Secretary of State for Oceans and International Environmental and Scientific Affairs.

Despite these amends and DOS's strong rhetorical acknowledgement of environmental issues, Wirth remained frustrated with the lethargy of the foreign policy bureaucracy in embracing new challenges and catering for them. In his 1997 official Earth Day remarks, he pointed to the problem of ever falling budgets, but ever increasing portfolios. An equally committed and equally frustrated Administrator of USAID, Brian J. Atwood, shared much of Wirth's criticism of the Clinton administrations' foreign policy establishment's dedication to environmental security. He summed up the frustrations pointedly: 'To look at the FY96 foreign affairs budget, one would have to conclude that many in the Congress believe that new challenges can still be addressed by old methods, or failing that, safely ignored'.[66] In December 1997 Wirth gave in to his frustration and resigned from his post as head of the OUSGA, and left government to

[64] Christopher cited in Thomas Lippman, 'Christopher puts Environment at Top of Diplomatic Agenda', *Washington Post*, 15 April 1996.
[65] Frank Clifford, 'Christopher Calls for Emphasis on Resources', *Los Angeles Times*, 10 April 1996.
[66] Brian J. Atwood, 'Remarks to the Conference on New Directions in US Foreign Policy at the University of Maryland, College Park', reproduced in

head the Washington-based United Nations Foundation, where he has remained since.

Wirth, had he not taken the decision to resign, may have been sacked that year anyway. According to his own admission, he was far too radical in environmental terms for anything the administration wanted to do. In his criticism, Wirth has no qualms about extending the bygone turf war to the White House and lashing out against his old friend Gore, who, according to Wirth, was only committed to the environment at the beginning of the administration. Once, however, Gore was set to run for President in the 2000 presidential elections – round about 1996 – his environmental focus crumbled and he went back on himself. Given the nature of Wirth's departure from office, comments such as this one cannot be evaluated without taking account of a certain bias on Wirth's part; nonetheless, it is important to note that Wirth is not alone in his conviction of Gore's turn away from environmentalism. Many committed environmentalists believe, for example, that Gore had no real interest in getting a stricter deal at Kyoto.[67] A 'bad deal' for the American economy (and by extension people) but a good deal for the environment would not have done Gore any favours during the election campaign. As Walter A. Rosenbaum pointedly put it, 'being green doesn't appear to help Democrats get to the White House'.[68]

Gore's followers, however, find such criticism both absurd and unfair. The argument here is of course that without Gore environmental issues would never have been so high profile in the first place. After all, it was Gore who created Wirth's office. Moreover, alongside the OUSGA it was Gore who was a strong proponent of environmental diplomacy. Altogether he co-chaired five bi-national commissions: the already mentioned Gore–Chernomyrdin Commission; the US–Kazakhstan Joint Commission; the US–Egypt Partnership for Economic Growth and Development; the US–South Africa Bi-national Commission; and the US–Ukraine Bilateral Commission,

Environmental Change and Security Project Report (Washington DC: The Woodrow Wilson Center, 1996), p. 86.

[67] Author interview with the Director of the Sierra Club's International Program Larry Williams and Program member Dr Robert Smythe, 17 September 2005, Georgetown, Washington DC.

[68] Walter A. Rosenbaum, *Environmental Politics and Policy*, fifth edition (Washington DC: Congressional Quarterly Press, 2002), p. 25.

all of which have a strong environmental focus that can be traced back to Gore. The environmental working group (EWG) under the Gore–Chernomyrdin Commission, for example, focused strongly on many of the ideas Gore already nourished during his time in the Senate, namely the use of intelligence capabilities for environmental research. One of the six environmental issue areas where intelligence was put to use was that of military base cleanup. The programmes required the former enemies to exchange formerly top-secret intelligence, whereby the first phase of the project focused on two military facilities in each of the two countries. According to Deputy Under Secretary of Defense for Environmental Security (1993–2000) Goodman the exchange took place as follows:

The first set of products was to focus on one site in each country, contaminated primarily with petroleum oils and lubricants or what's known as 'POL'. The US delegation has prepared a derived product on Eysk Air Force Base, near the Okhotsk Sea. [...] And, the Russian delegation prepared a derived product map on Eglin Air Force Base, in Florida. They are both time-lapse analyses from the 1970s to the present. [...] The derived products illustrate locations and types of contamination at each military site, over the last 20 years. And, they also indicate possible pathways for contamination which could affect human, animal, or plant life in the surrounding region.[69]

And, again according to Goodman, with the following benefits:

[The] exchange is important for several reasons. First, from an environmental perspective, the use of the classified intelligence assets may help us save time, money, and manpower in identifying types and extent of environmental contamination and in providing the risk assessment based on the location of the pathways and the receptors being humans or animals. [...] Second, from a more general perspective of US/Russian relations, this product is an outstanding example of establishing contacts and developing trust between two groups of people who have long been taught not to trust each other – members of the intelligence communities. It has also, I might add, helped develop [...] the defense and environmental communities in both our countries, which, I believe, is particularly valuable. [...] the sharing of such information between these two countries, that have

---

[69] Goodman in Ashton Carter, 'DoD News Briefing on the New Information Sharing Initiative with the Russian Government, a Result of the Recent Gore–Chernomyrdin Commission Meeting' (Washington DC: Office of the Assistant Secretary of Defense, Public Affairs, 1996), at www.defenselink.mil/transcripts/transcript.aspx?transcriptid = 275 [1/2005].

valuable intelligence assets under their command, raises the possibility of sharing this information with other countries as well. As such, we will jointly be able to contribute to enhanced environmental knowledge throughout the globe.[70]

In short, the efforts of the EWG were a prime example of the dual use of intelligence. Representative for all participants, the Administrator of the National Oceanic and Atmospheric Administration, Jim Baker, summed up the mood towards Gore pointedly as follows: 'all of this, [the programme] goes back to the vision of the Vice President, who saw the opportunity to do something that couldn't be done before this linking'.[71]

As regards the use of intelligence gathering equipment in general, Gore's role cannot be overemphasised. After all, he was one of the first who linked environmental issues with intelligence capabilities, and it was he who played a main part in the creation of SERDP in 1990, and in that of MEDEA in 1995, the project whereby some sixty scientists received security clearance and permission to programme US spy satellites to study a dozen environmentally sensitive areas.[72] Never before had civilians been allowed access to top secret intelligence equipment; indeed during the Cold War the usage of intelligence equipments for civilian research purposes was simply unthinkable. Driven by the input of Gore, who believed that the intelligence community could be a vital asset in deepening research into global warming, the intelligence community slowly became acquainted with the idea, and under MEDEA spy satellites were used to, for example, examine the effects of the rising carbon dioxide levels on Tanzania's Mount Kilimanjaro's high forests.

Why did they securitise?

So far this chapter has shown that the Clinton administrations' classification of environmental problems as existential threats resulted in a change of behaviour on the part of the securitising actor in such a

---

[70] *Ibid.*
[71] *Ibid.*
[72] According to Dabelko some claim that MEDEA is an acronym for Measurement of Earth Data for Environmental Analysis while others are quoted as saying it is not an acronym (Dabelko, 'Tactical Victories and Strategic Losses', p. vii).

way that this case study constitutes a successful securitisation. This change of behaviour manifested itself in terms of leadership, commitment, budget and in the creation of new policies and/or institutions that dealt with issues of environmental security. In more detail, these were the creation of two new offices concerned with environmental security at Under Secretary of State and Deputy Under Secretary of Defense level respectively; the creation of several international environmental security programmes, including AMEC and the MOU; the creation of twelve regional environmental hubs; the creation of environmental diplomacy, for example in form of the environmental working group under the Gore–Chernomyrdin Commission; the creation of MEDEA and the creation of the State Failure Task Force. In other words, the Clinton administrations had – if considerably varying in size and budget – active environmental security programmes both in the domestic and in the international realm. What remains is to answer the question: *why* did they do this? In the theoretical framework developed in Chapter 2 it was suggested that the answer to this question is inseparably connected to the actual beneficiary of the securitisation. It was further suggested that (at least) two such beneficiaries are possible. The first is the referent object of security as identified by the securitising actor; the second is the securitising actor. I further argued that an 'agent benefiting securitisation' is apparent when a securitisation is or was inconsistent in its own terms. That is, when there is or was a discrepancy between the saying (existential threat justification) and the doing (security practice). Therefore, in order to answer the above question, we need to look at whether or not the Clinton administrations' environmental security policies and practices were consistent with the threats they identified.

Judging by the many official statements by the Clinton administrations – including the various National Security Strategies – threats to do with the environment were identified as acute, omnipresent and, above all, as originating from a globally deteriorating environment, including global climate change and ozone depletion. Threats were identified as knowing no national boundaries, with environmental problems overseas believed to directly affect the United States' national security. If this was how environmental threats were perceived, however, then it is reasonable to suggest that the Clinton administrations' environmental security efforts were *not* consistent with the threats they identified. With an annual budget of $5 billion the ODUSD–ES was by far the biggest player in US environmental security. However, this budget was not used to avert one or more of the stated environmental threats, but rather to enable the military to clean up its own act and comply better with environmental legislation in future. In the resulting 'greening of the military', the environmental security equation was not made in relation to global environmental change and its anticipated consequences, but rather it was made in so far as the ill-functioning of the environment posed a threat to military readiness and, consequently, to the military's ability to provide national security.[73] If the Clinton administrations would have wanted to address global environmental change then, at the very least, they would have needed to embrace conservation, cleanup and pollution prevention *nationwide*, instead of the confinement of these important environmental initiatives to military land only. Similarly, they would have needed to promote prevention of pollution everywhere, rather than only military pollution, particularly given that industrial pollution accounted for the largest amount of national pollution in the United States at that time. Education and training in environmental issues would have to have included the entire public, as only then could a deep understanding of the problems of environmental degradation and their prevention – such as a change in consumption patterns – have been achieved. The 1994 and 1995 National Security Strategy for one suggests as much.

Domestically, the US must work hard to halt local and cross-border environmental degradation. In addition, the US should foster environmental technology targeting pollution prevention, control, and cleanup. Companies that invest in energy efficiency, clean manufacturing, and environmental services today will create the high-quality, high-wage jobs of tomorrow. By providing access to these types of technologies, our exports can also provide the means for other nations to achieve environmentally sustainable economic growth. At the same time, we are taking ambitious steps at home to better manage our natural resources and reduce energy and other consumption, decrease waste generation and increase our recycling efforts.[74]

---

[73] On the inconsistency between the NSS and DOD action, see also Jon Barnett, *The Meaning of Environmental Security: Ecological Politics and Policy in the New Security Era* (London: Zed Books, 2001), p. 79.

[74] '1994 and 1995 US National Security Strategy of Engagement and Enlargement', extract reproduced in *Environmental Change and Security Project Report* (Washington DC: The Woodrow Wilson Center, 1995), p. 49.

Considering this discrepancy between what the Clinton administrations said about environmental security and what was actually
done in the name of such a policy, the beneficiary of their environmental security practices and policies cannot possibly have been
the American people. If this is so, something in its place must have
been the beneficiary of these policies. I propose that this something
was the national security establishment, and consequently that this
case study is an example of agent benefiting securitisation. This suggestion is substantiated not only by what was done in the name of
environmental security (for example, base cleanup to ensure military
readiness or pollution prevention to avoid fines), but also by the acute
need for the discourses of danger in the immediate post-Cold
War period.[75] Thus at that time, security agencies (DOD, DOE,
CIA etc.) that were threatened in their survival (at least as they were
then) branched out to incorporate many comparatively small discourses of danger (including environmental security, economic security, rogue states etc.) to fill the void left by the disintegration of the
East–West conflict, in the process constructing a security threat, whilst
at the same time offering their 'services' as providers of (environmental) security. The fact that environmental security was attractive
for governmental agencies because it gave them a raison d'être
and ensured continuous federal funding is powerfully evinced by the
international environmental security efforts, especially AMEC and
those under the MOU, which were popular precisely because agencies
(such as the EPA and the DOE) hoped to end notorious under-funding
through participating in these programmes. Likewise, in his analysis
of the use of intelligence capabilities for environmental ends, the
Canadian political scientist Ronald J. Deibert even speaks of a
'military environmental security complex'. He argues:

Once environmental causes of military conflict were identified, however,
a potential new 'enemy' was introduced into the security scheme of
things. Image analysts accustomed to identifying the latest Soviet submarines would now have to monitor the depletion of fresh water resources
in 'strategic' areas. [...] For the large aerospace corporations and their
employees that had once thrived on a steady stream of Cold War defence

---

[75] I am not the first to link the emergence of US environmental security specifically
with Campbell's discourses of danger: see Barnett, *The Meaning of
Environmental Security*, pp. 48, 88.

contractors, it meant new business and the resurrection of jobs seemingly doomed by the loss of an enemy. For the military, while it wasn't the Gulf War, it *was* a mission, and missions were getting harder to find in a 'world of uncertainty'.[76]

It is vital to note here that Deibert's analysis is not simply an academic critique of the intelligence community. Instead, both the fact that the intelligence community was looking for relevancy in the post-Cold War era, and also that the possibility of environmentally induced conflict was seen as a legitimate cause to focus on environmental security, was confirmed to me by the Clinton administration's Deputy National Intelligence Officer for Global and Multilateral Affairs Richard Smith in an interview for this book.[77] This particular position was created in late 1993 and the environment was one of twenty-two 'enduring challenges' that occupied the Intelligence Community.[78]

Another example that supports the claim that this case study is one of agent benefiting securitisation is the experience of the Under Secretary of State for Global Affairs Wirth, who sought to broaden his office's environmental security agenda in order to address the problems of environmental insecurity the administration itself had identified, only to resign from his job in anticipation of being made redundant for his efforts.

Considering all that has been said here, it should now be clear that there are almost as many reasons why the Clinton administrations securitised the environment as there are agencies in the national security establishment. Regardless of the various ways in which this would happen, each agency saw environmental security as an opportunity to benefit themselves. My conclusion here then is that in addition to the encouragement provided by the end of the Cold War, such self-serving behaviour was readily facilitated by the indeterminacy of the concept of environmental security.[79]

---

[76] Ronald J. Deibert, 'From Deep Black to Green? Military Monitoring of the Environment', *Environmental Change and Security Project Report* (Washington DC: The Woodrow Wilson Center, 1996), p. 29.

[77] Author telephone interview with Deputy National Intelligence Officer for Global and Multilateral Affairs Richard Smith, 15 May 2008.

[78] Richard Smith, 'The Intelligence Community and the Environment: Capabilities and Future Missions', *Environmental Change and Security Project Report* (Washington DC: The Woodrow Wilson Center, 1996), p. 106.

[79] Compare, Dabelko, 'Tactical Victories and Strategic Losses', p. 111.

Conclusion

In the theoretical framework of this book I argued that by identifying the beneficiary of securitisation we can gain insights into the intentions of the securitising actor. The analysis of this chapter supports this theoretical claim. Thus once the beneficiary of US environmental security is identified as having been the national security establishment and not the stated American people, it is obvious that this holds implications for the intentions of the securitising actor. Clearly the intentions for why an actor securitised cannot simply be read off from what an actor says. This case study confirms the theoretical claim made earlier in this book – that actors do not always and necessarily act with a view to securing the stated referent object of security. Instead, securitising actors sometimes securitise because of the benefits such action has for themselves; a state of affairs that has here been called agent benefiting securitisation.

Given that US environmental security under the Clinton administrations was an agent benefiting securitisation, this case study raises important questions about the moral worth of environmental security. Two pivotal questions in this context are: who or what should benefit from an environmental security policy? and when, if ever, is a securitisation in the environmental sector morally right? These questions will be answered in Chapter 6. For now, it will be necessary to look at the fate of environmental security as of 2001, when the first Bush administration came to power.

5 | The Bush administrations and environmental security

Introduction

This chapter examines whether the environment remained a security issue under the Bush administrations or whether there was a shift in policy-making which led to the desecuritisation of the environment. It is possible to begin by taking it for granted that the environment was securitised when the first Bush administration came into office, as every new President inadvertently 'inherits his predecessor's budget, structures, commitments, and bureaucratic personnel' upon taking office.[1] This chapter will examine what happened to all the existing domestic and international environmental security initiatives and programmes, as outlined in the previous chapter. Did they remain in place? Did they change? If yes, in what way? These are just some of the questions addressed.

First, it will be necessary to examine the first Bush administration's national security strategy in light of the historical context in which it took place. For the first Clinton administration the historical context was the end of the Cold War, for the first Bush administration it was the events of 11 September 2001 and the resultant 'war on terror'. Although the first Bush administration was in office for ten months prior to 9/11, these events shaped national security the most, with the first NSS by the administration issued on 17 September 2002, a year into the ongoing war on terror.

From its outset the first Bush administration had a different attitude to environmental security than the Clinton administrations, ending programmes wherever they could, and changing the more institutionalised initiatives/programmes like the Office of the Deputy Under Secretary of Defense, Environmental Security and calling this particular office the Office of the Deputy Under Secretary of Defense,

---

[1] Sam C. Sarkesian, John Williams, John Allan and Stephen Cimbala, *US National Security: Policymakers, Processes and Politics*, third edition (Boulder: Lynne Rienner, 2002), p. 96.

Installations and Environment (ODUSD–I&E). Although this may suggest a straightforward desecuritisation of the environment, it is not that simple. For example, although the ODUSD–ES was renamed and no longer existed in its original form, many of the initiatives essential to defence environmental security, such as pollution prevention and compliance, remained in place. In addition, some international environmental security programmes, such as AMEC, were still in place; and in the case of AMEC with the admission of the United Kingdom as a new member only in 2004, thriving. Moreover, again it must be remembered that for a securitisation to be in place the use of the word 'security' is not required, as it is the *grammar of security* that makes an issue a security issue, not the application of a label. In other words, although at first sight there seems to be little doubt that the Bush administrations desecuritised the environment, a more rigorous analysis of the issues at stake here is required. Indeed, before jumping to conclusions it must be noted that all new administrations, particularly after a party political swing as in the Clinton–Bush case, are eager to change labels, names, structures, frankly anything that can be changed; so that on the surface nothing resembles the administration of the predecessor.

Setting the historical context: the first Bush administration and the 'war on terror'

When the Bush administration came into power in early 2001 they neither had a very coherent foreign policy-making agenda nor a very coherent national security policy.[2] Initial acts of unilateralism rebuffed friends and sympathisers, and President Bush's already meagre public approval rating was soon on an ever downward sliding scale.[3] The events of 11 September 2001 changed all of this. Suddenly the Bush administration had a consistent foreign and security policy, with the two becoming once again synonymous.[4] At the same time, President

---

[2] Stefan Halper and Jonathan Clarke, *America Alone: The Neo-Conservatives and the Global Order* (Cambridge University Press, 2004), p. 131.
[3] Jane Martinson, 'Poll Shows Half of Americans Doubt Bush's Trustworthiness. Special Report on George Bush's America', *The Guardian*, 28 May 2001.
[4] Sarkesian *et al.*, *US National Security,* p. 14; Ronald Asmus, 'The European Security Agenda', in Roland Dannreuther and John Peterson (eds.), *Security Strategy and Transatlantic Relations* (Abingdon: Routledge, 2006), pp. 18–19.

Bush's approval ratings skyrocketed to 92 per cent, the highest approval rating of any President ever.[5]

For national security the events of 9/11 had three obvious and related repercussions. The first was the beginning of the so-called 'war on terror'. The second was the establishment of the Department of Homeland Security, created largely to help with the war on terror at home. The third was an increase in defence spending to finance the war on terror. During the period between FY 2001 and FY 2008 the defence budget increased by 62 per cent to a total of $481.4 billion.[6] President Bush's much-cited State of the Union address of January 2002, which contained the infamous 'axis of evil' reference, captured this newfound direction to national security well.

My budget includes the largest increase in defense spending in two decades – because while the price of freedom and security is high, it is never too high. Whatever it costs to defend our country, we will pay. The next priority of my budget is to do everything possible to protect our citizens and strengthen our nation against the ongoing threat of another attack. [...] My budget nearly doubles funding for a sustained strategy of homeland security, focused on four key areas: bioterrorism, emergency response, airport and border security, and improved intelligence.[7]

Besides the war on terror and all that it includes, Bush equated only one further issue with the status of national security – the economy. Hierarchically, however, the economy was subordinate to the war effort – 'once we have funded our national security and our homeland security, [then] the final great priority of my budget is economic security for the American people'.[8]

In the remainder of the speech, other issues, such as health and retirement, were also affixed with the label 'security'; this time, however, not as a national security sentiment but in a social security way.

---

[5] Gary Langer, 'Still Proud to be an American – Poll: One Year Later Public Remains Proud, Optimistic despite Fears', *ABC News*, 10 September 2002.

[6] Office of Management and Budget, 'Budget of United States Government FY2008' (Washington DC: Department of Defense, 2008) at www.whitehouse.gov/omb/budget/fy2008/defense.html (3/2009).

[7] George W. Bush, 'State of the Union', 29 January 2002, at http://georgewbush-whitehouse.archives.gov/news/releases/2002/01/20020129–11.html [6/2006].

[8] *Ibid.*

In all of the speech, the environment was mentioned only in one sentence, without any link to security, but rather as an important domestic policy-making issue.

Members, you and I will work together in the months ahead on other issues: productive farm policy, a cleaner environment, broader home ownership, especially among minorities, and ways to encourage the good work of charities and faith-based groups. I ask you to join me on these important domestic issues in the same spirit of cooperation we've applied to our war against terrorism.[9]

The National Security Strategy published later that year in September 2002 was on par with the State of the Union address in every aspect. The war on terror loomed large, whilst the environment was no longer mentioned with reference to security. In fact, the environment was only mentioned twice in the entire thirty-plus-page document. Only one paragraph, in a section entitled 'Ignite a new era of global economic growth through free markets and free trade', showed some recognition of environmental diplomacy. It read:

Protect the environment and workers. The United States must foster economic growth in ways that will provide a better life along with widening prosperity. We will incorporate labor and environmental concerns into US trade negotiations, creating a healthy 'network' between multilateral environmental agreements with the WTO, and use the International Labor Organization, trade preference programs, and trade talks to improve working conditions in conjunction with freer trade.[10]

The environment was mentioned a second time as part of a section entitled 'Expand the circle of development by opening societies and building the infrastructure of democracy'.

The United States seeks a constructive relationship with a changing China. We already cooperate well where our interests overlap, including the current war on terrorism and in promoting stability on the Korean peninsula. Likewise, we have coordinated on the future of Afghanistan and have initiated a comprehensive dialogue on counterterrorism and similar transitional concerns. Shared health and environmental threats,

[9] *Ibid.*
[10] National Security Strategy of the United States of America, September 2002, available at http://georgewbush-whitehouse.archives.gov/nsc/nss/2002/index.html.

such as the spread of HIV/AIDS, challenge us to promote jointly the welfare of our citizens.[11]

Notably, although here 'environmental threats' were mentioned, no examples of such threats were provided, nor strategies outlined as regards how to alleviative such threats.

Although the level of concern for the environment in these two important documents was in stark contrast to that of the Clinton administrations, the Bush administration's negligence of environmental issues did not come as a surprise. From the outset – and indeed as Governor of Texas – President Bush showed little interest in environmental issues, and amongst some of his first actions in government was the roll-back of many of the environmental rules and regulations put in place by his predecessor.[12]

Bush laid out his environmental intentions within 90 minutes of being sworn in, when his chief of staff, Andrew Card (formerly the top lobbyist for General Motors) sent out a memorandum putting 371 of Clinton's pending [environmental] rules on hold. [...] By mid-March 2001, Bush [...] started throwing bombs of his own. He retreated from his campaign promise to regulate carbon dioxide (the primary contributor to global warming), announced that he would consider all public lands fair game for oil and gas drilling, and delayed application of the Roadless Area Conservation Rule, which protected 58 million acres of public lands from logging, mining, and development.[13]

This said, it is important to note here that President Clinton's environmental policy-making record was far from perfect either; many environmental rules and regulations were last-minute efforts, put in place only during the 'lame-duck days' before President Bush's inauguration.[14] This book, however, is not concerned with environmental

---

[11] *Ibid*. p. 27.
[12] When the future President Bush was Governor, Texas had the worst environmental record of any state in the US. Amongst his worst bits of environmental deregulation was his permissive approach to 'grandfathered' coal-fired plants (established long before the regulations enacted by the 1977 Clean Air Act) that were given a voluntary choice to end production if they wished to do so, leading to many plants remaining active as normal.
[13] Carl Pope and Paul Rauber, *Strategic Ignorance: Why the Bush Administration is Recklessly Destroying a Century of Environmental Progress* (San Francisco: Sierra Club Books, 2004), pp. 45–6.
[14] *Ibid*. p. 45.

policy-making per se and so it will not analyse the nature of the different administrations' environmental records here in any more detail. Nonetheless, in view of what has been said here about Bush's environmental record, it is realistic to assume that even without the events of 9/11, environmental security would not have been part of the Bush administration's national security strategy, simply because they were not environmentalists. And in the absence of a commitment to protecting the environment, it is hard to imagine that this administration would have been a strong advocate of environmental security even without the war on terror. After all, and as the previous chapter has found, leadership was one of the most important catalysts in the realisation of environmental security.

What is more, evidence suggests that even before the events of 9/11 the Bush administration began to actively dismantle parts of Clinton's environmental security strategy, providing strong grounds for the suggestion that 9/11 was merely a facilitator and not the reason for going back on the predecessor's environmental security policies. It is also of note that the same applies for the increase of military spending. During his election campaign 'George W. Bush warned that US defense spending was failing to keep pace with US defense needs and missions'.[15] Only a few weeks into the administration President Bush ordered a comprehensive review of the military to be conducted by Andrew Marshall, the neo-conservative long-term head of the Pentagon's think tank, the Office of Net Assessment, who was renowned for having had a long-term interest in military transformation.[16] In short, the 'decision to rebuild the armed forces came *before* 9/11, and was not a response to the security threat'.[17] In the first instance, then, 9/11 was a facilitator for increased military spending, not the reason. Likewise, as Stuart Croft notes, the 'meaning [of the events of 9/11] did not come "out of the blue" [...] terrorism had been a major theme in government and in popular culture

[15] Charles W. Kegley, Eugene R. Wittkopf and James M. Scott, *American Foreign Policy*, sixth edition (London: Thomson Wadsworth, 2003), p. 80.
[16] Nicholas Lemann, 'Dreaming about War', *The New Yorker*, 16 July 2001; Douglas McGray, 'The Marshall Plan', *Wired* (2003) at www.wired.com/wired/archive/11.02/marshall.html [2/2008].
[17] Dan Meyer and Volk E. Everett, 'W for War or Wedge? Environmental Enforcement and the Sacrifice of American Security – National and Environmental – to Complete the Emergence of a New "Beltway" Elite', *Western New England Law Review* 25 (2003), p. 50 (emphasis added).

before'[18] – a finding that supports Campbell's assertion that danger is always crystallised around more than one referent object, even if not all of them are visible at all times.[19]

Environmental security domestic programmes and initiatives

Department of Defense

Under the Clinton administrations the domestic environmental security programme of the US was made up entirely by the efforts of the Department of Defense, or more precisely by those of the ODUSD–ES. Since no new offices/initiatives for the provision of domestic environmental security were put in place by the Bush administrations, this section will focus entirely on the DOD and on what remained of its defence environmental security programme.

One obvious change to the Clinton-era DOD environmental security programme was the name change away from ODUSD–ES to the Office of the Deputy Under Secretary of Defense, Installations and Environment (ODUSD–I&E). Since the label 'security' in itself is not imperative for securitisation,[20] the renaming might not be of interest here, were it not for what exactly the office was renamed into, or rather made part of. Thus the Bush administrations' DOD office in charge of environmental issues did not stand alone, but rather had been grouped together with the Office of Installations, a wealthy office in charge of the oversight of billions of dollars' worth of military installations.[21] This reorganisation, according to Professor of Public Administration and Policy Robert F. Durant, signalled 'the demise of environmental security as a central component of national defense policy [and] it also failed to put a fence around ENR [environmental and natural resources] funds, thus leaving

---

[18] Stuart Croft, *Culture, Crisis and America's War on Terror* (Cambridge University Press, 2006), p. 266.

[19] David Campbell, *Writing Security*, second edition (Minneapolis: University of Minnesota Press, 1998), p. 171.

[20] Ole Wæver, *Concepts of Security* (Copenhagen: Institute of Political Science, University of Copenhagen, 1997), p. 49.

[21] Under Clinton there was a stand-alone Deputy Under Secretary Office for Industrial Affairs and Installations, incidentally directed by Sherri W. Goodman's husband John B. Goodman.

them susceptible to budget raids for core military responsibilities on installations'.[22] The following job description, of the 2004 inaugurated Deputy Under Secretary of Defense for Installations and Environment Phillip W. Grone, highlights the multitasking of this office well.

Mr. Grone has management and oversight responsibilities for military installations worldwide, which have a land area covering over 46,000-square miles and containing 587,000 buildings and structures valued at more then $640 billion. His responsibilities include the development of installations capabilities, programs, and budgets; base realignment and closure; privatization of military housing and utilities system; competitive sourcing; and integrating installations and environment needs into the weapons acquisition process. Additionally, he has responsibility for environmental management, safety and occupational health; environmental restoration at active and closing bases; conservation of natural and cultural resources; pollution prevention; environmental research and technology; fire protection; and explosives safety.[23]

Given the scope of the installations side of this office, it remains to be seen how important the environmental element was and whether it had been sidelined in favour of military installations. It is worthy of note here that all the 2001 Defense Environmental Quality (EQ) annual report to Congress issued by the ODUSD–I&E had to say about the reorganisation was that 'it will help DOD's environmental programs to be more effective'.[24] How so, was not specified. My own enquiries regarding the name change came to nothing; not a single person in the new office could tell me the reasons, while other people interviewed for the purposes of this book argued that the

[22] Robert F. Durant, *The Greening of the US Military: Environmental Policy, National Security and Organizational Change* (Washington DC: Georgetown University Press, 2007), p. 228.

[23] The author downloaded this statement from the ODUSD–I&E website in 2006. Mr Grone left the DOD in mid-December 2007 for the private sector, and with his departure the statement was removed from the website. Alex Beehler served as Acting Deputy Under Secretary from mid-December to mid-February 2008. Wayne Arny was appointed Deputy Under Secretary on 15 February 2008.

[24] Office of the Deputy Under Secretary of Defense, Installations and Environment, '2001 Department of Defense Environmental Quality (EQ) Program Annual Report to Congress' (Washington DC: Department of Defense, 2001), p. 4.

Table 2 *BRAC budget summary 2001–2009 (in million US dollars)*[25]

FY	2001	2002	2003	2004	2005	2006	2007	2008	2009 appropriated
Total	801	625	771	387	250	569	497	527	524

ODUSD–ES was considered as very closely affiliated with the Clinton administrations and that it had to go as part of an 'out with the old, in with the new' strategy.

Moving on from the name change to the actual environmental responsibilities of the new office, these were less with the cleanup of military bases, compliance with federal, state and international environmental law, pollution prevention, and conservation of some of the same environmental programmes that were in existence under Clinton. Moreover, as before, environmental cleanup remained the single biggest environmental issue funded by the DOD. Unlike under the Clinton administrations, however, there was no longer a stable annual $5 billion budget, out of which about half was allocated for cleanup activities each year. Instead, the so-called Defense Environmental Restoration Program (DERP) was funded from two separate accounts, the Environmental Restoration (ER) account and the Base Realignment and Closure (BRAC) account.[26] Whilst funding levels for ER programmes have remained relatively stable over the past decade (averaging $1.3 billion annually), funding for BRAC has fluctuated.[27] For details of the latter, view Table 2 above.

---

[25] The data for FY 2001 is taken from Defense Environmental Programs (DEP) Report to Congress for FY 2004; data for 2002–2004 taken from Defense Environmental Program (DEP) Report to Congress FY 2005; data for 2005 taken from Defense Environmental Program Report to Congress FY 2006; data for 2006–2009 taken from Defense Environmental Programs (DEP) Report to Congress FY 2007; data for 2007–2009 taken from Defense Environmental Programs Report to Congress FY 2008. All reports are available at www.denix.osd.mil/portal/page/portal/denix/environment/ARC.

[26] Office of the Deputy Under Secretary of Defense (Installations and Environment), '2004 Defense Environmental Programs (DEP) Annual Report to Congress' (Washington DC: Department of Defense, 2004), p. 2.

[27] *Ibid.* p. 2.

According to the ODUSD–I&E, however, this fluctuation in funding is a sign of success, as money increases or decreases according to demand. The decrease in funding (a more than halving between FY 2001 and FY 2005) thus signified that the DOD had completed many of the ER requirements outlined under BRAC.[28] The rise in funding from FY 2006 onwards, in turn, did not equate to the failure of the programme, but rather during those years the BRAC commission – or its ambit – grew/changed extensively. Thus, with over 800 'distinct and identifiable recommended BRAC "close" or "realign" actions [...] the 2005 BRAC recommendations exceeded the number considered by all prior BRAC Commissions [in 1988, 1991, 1993 and 1995] combined'.[29] A possible explanation of these occurrences can be found in a changed focus of BRAC away from closing bases, towards accommodating 'the new security needs [i.e. the war on terror] where many military capabilities are surging to meet new battlefield requirements'.[30] The purpose of BRAC then was 'to bring our nation's military infrastructure into line with the needs of our armed forces, not only by reducing costs and closing un-needed installations, but also by facilitating the transformation of our armed forces to meet the challenges of the new century'.[31] The 2005 BRAC commission report found that:

In previous rounds of BRAC, the explicit goal was to save money and *downsize* the military in order to reap a 'peace dividend'. It was clear from the Commission's examination of the DOD 2005 BRAC list that the historical goal of achieving savings through eliminating excess capacity was *not* always the primary consideration for many recommendations. In fact, several DOD witnesses at Commission hearings made it clear that *the purpose of many 2005 BRAC recommendations was to advance the goals of transformation, improve capabilities, and enhance military value.*[32]

Given these new security goals, it was not surprising that more BRAC cases were being reported (i.e. realignment was needed in order to accommodate the war on terror), whilst fewer bases were being put

---

[28] ODUSD–I&E, '2004 Defense Environmental Programs (DEP) Annual Report to Congress', p. 3.
[29] Defense Base Closure and Realignment Commission 2005, 'Final Report to the President', p. iii at www.brac.gov/finalreport.html [12/2005].
[30] *Ibid.* p. 317.
[31] *Ibid.* p. 1.
[32] *Ibid.* p. 3 (emphases added).

forward for closure. In other words, more money for BRAC did not automatically mean more money for environmental cleanup, especially considering that environmental cleanup is of particular importance when bases are being closed and returned to the public realm and not when they are being reformed to accommodate new security threats. The environment received little mention in the 2005 report anyway. Whereas the previous BRAC report from 1995 stated that 'the environmental impact' in *all* BRAC projects had to be considered,[33] the 2005 criteria required the Department of Defense to consider 'the impact of *costs* related to potential environmental restorations, waste management and environmental compliance activities',[34] indicating that environmental concerns were tied to whatever costs they may induce, whereas in previous BRAC rounds they were considered irrespective of such costs.

Unlike BRAC, environmental restoration (cleanup) was not subject to the national security and defence requirements, and thus remained stable irrespective of the change in national security. As the previous chapter has shown, the DOD has been legally required to clean up its bases since 1975 and there is little any one administration can do to change this. Similarly, environmental compliance – achieved in part by the preferred method of pollution prevention – and conservation were all fixed legal requirements, laid down in numerous federal, state and international laws. Given the existence of environmental laws, all administrations, whatever the nature of their environmental commitment, are bound to obey these laws, and funding must be made available for these issues. The same applied therefore to the Bush administrations with their dubious commitment to environmental issues. Table 3 sets out in detail the Environmental Quality budget from 2001 to 2009 – made up of compliance, pollution prevention and conservation – under the Bush administration.

Especially when compared with Table 1 (page 91), this table shows that with the advent of the Bush administration the budget for both compliance and pollution prevention decreased – in the case of

---

[33] Defense Base Closure and Realignment Commission 1995, 'Report to the President' (Washington DC: Department of Defense, 1995), p. x at www.brac.gov/ShowPubDoc.aspx?Doc_st=317200624247_1.pdf&Path_st=1995\BRCM&DocID_in=1648 [12/2005].

[34] Defense Base Closure and Realignment Commission 2005, 'Final Report to the President', p. 317.

Table 3 *Environmental Quality budget 2001–2009 (in US dollars)*[35]

Year	Total budget	Compliance	Pollution prevention	Conservation
2001	2 billion	1.63 billion	212 million	183 million
2002	2 billion	1.67 billion	226 million	157 million
2003	2.2 billion	1.80 billion	187 million	179 million
2004	1.9 billion	1.66 billion	116 million	158 million
2005	2.0 billion	1.67 billion	125 million	188 million
2006	1.8 billion	1.54 billion	125 million	204 million
2007	1.86 billion	1.43 billion	130 million	300 million
2008	1.97 billion	1.49 billion	121 million	352 million
2009 (appropriated)	2.08 billion	1.67 billion	165 million	344 million

compliance by as much as $660 million, from the height of funding of $2.2 billion for FY 1996 to just $1.54 billion ten years later in FY 2006. In the case of pollution prevention, the decline in budget was even more dramatic: in the dozen years between FY 1994 and FY 2006 the annual budget declined by 63 per cent, from $338 million to $125 million. The conservation budget on the other hand increased steadily, with the appropriated budget for FY 2009 ($330 million) more than three times that for FY 1994 ($99 million). In the present case study, however, funding levels alone, precisely because the DOD is by law required to do something about the environmental issues specified, are not an adequate indicator of the status of environmental security. They need to be analysed with regard to what was being funded and why; only then can they provide useful insights into the prevailing trends of the administrations' policy-making direction.

The tone for the Bush administrations' defence environmental policy was set in March 2002 when a distinctly different message regarding environmental issues was first heard from the DOD. Unlike

[35] Data from FY 2001–04 taken from the Office of the Deputy Under Secretary of Defense, Installations and Environment, '2005 Defense Environmental Programs (DEP) Annual Report to Congress', p. C-2; data for 2005 taken from the 2006 Defense Environmental Programs (DEP) Annual Report to Congress; data for FY 2006 taken from the 2007 Defense Environmental Programs (DEP) Report to Congress; data for FY 2007–2009 taken from the 2008 Defense Environmental Programs (DEP) Annual Report to Congress. All reports are available at www.denix.osd.mil/portal/page/portal/denix/environment/ARC.

under the Clinton administrations, talk was no longer about the need
to bring in or foster environmental security. Instead, the argument was
that military security must not be jeopardised in order to comply with
environmental requirements, the same ones that were formerly called
'environmental security'. A policy draft leaked to the policy watchdog
Public Employees for Environmental Responsibility (PEER) cites
the 2002 draft copy of the Sustainable Defense Readiness and
Environmental Protection Act (SDREPA) as follows:

Federal departments and agencies shall not place the conservation of public
lands, or the preservation or recovery of endangered, threatened, or other
protected species found on military lands, above the need to ensure that
soldiers, sailors, airmen and Marines receive the greatest possible preparation for, and protection from, the hazards and rigor of combat through
realistic training on military lands and in military airspace.[36]

In short, military security and wildlife protection were suddenly at
odds with each other, despite the fact that under Clinton the two had
been compatible for years. Where did this new direction come from?
And, what was the thinking behind it? According to staff within the
ODUSD–I&E the logic behind this 'sudden' incompatibility stemmed
from a strong prevalence of so-called encroachment issues. Encroachment issues, ODUSD–I&E staff were eager to stress, did start already
during the final year of Sherri W. Goodman's period in office, when
military officials, in one staffer's words, 'flocked' to the ODUSD–ES to
complain about the impossibility of training properly.[37] All the same,
critics maintain that 'after the September 11 terrorist attacks, the
military and its congressional allies kicked their anti-encroachment
offensive into even higher gear'.[38]

  Altogether there were eight encroachment issues identified by the
DOD, namely: (1) endangered species habitat on military installations; (2) unexploded ordnance and munitions constituents; (3) competition for radio frequency spectrum; (4) protected marine resources;

36 Department of Defense, 'Sustainable Defense Readiness and Environmental
   Protection Act Discussion Draft', discussion draft not for release, 7 March
   2003, p. 2, at www.peer.org/docs/dod/3-7-02_draft_DOD.pdf [1/2006].
37 Author interview with the Director for Environmental Readiness and Safety
   Office of the Secretary of Defense Curtis Bowling 15 September 2005, Office
   of the Deputy Under Secretary of Defense (Installations & Environment),
   Arlington VA.
38 Durant, *The Greening of the US Military*, p. 229.

(5) competition for airspace; (6) competition for air pollution; (7) competition for noise pollution; and (8) urban growth around military installations.[39] According to a glossy brochure entitled *Department of Defense Sustainable Ranges: Better Planning through Partnership*, which highlights in particular the eighth encroachment issue – urban growth around military installations – military training and testing is affected in the following ways: (1) limitations on night exercises; (2) noise restrictions that create unrealistic flight training operations; (3) prohibitions against digging foxholes, using smoke-screens, or operating vehicles that generate dust; (4) curtailed weapons practice; (5) artificial, fragmented manoeuvres by small units due to land-use restrictions or habitat avoidance rules.[40]

As a solution to ease encroachment issues, the DOD identified seven different pathways: policy; programming; leadership and organisation; legislation and regulation; outreach and engagement; compatible land use and buffering; and comprehensive reporting to Congress. Out of these seven, 'compatible land use and buffering' had with the Readiness and Environmental Protection Initiative (REPI) a separate funding account. REPI was concerned with the easement of environmental concerns within buffer zones. For FY 2005 the DOD requested some $20 million from Congress and was awarded $12.5 million. For 2006 the request was $20 million again, but $37 million was awarded to this initiative, and by FY 2007 that had risen to $40 million. Notably, this initiative also accounts for a considerable part of the increase in funding in the conservation budget from FY 2004 to FY 2005 and again to FY 2006.[41]

Besides 'compatible land use and buffering', 'outreach and engagement' was the second noteworthy pathway to ease encroachment. The

---

[39] United States General Accounting Office, *Military Lacks a Comprehensive Plan to Manage Encroachment on Training Ranges* (02-614) (Washington DC: US General Accounting Office, 2002), p. 1 at www.gao.gov/new.items/d02614.pdf [3/2006].

[40] Office of the Deputy Under Secretary of Defense, Installations and Environment, *Department of Defense Sustainable Ranges: Better Planning through Partnership* (Washington DC: Department of Defense, 2005), p. 9.

[41] Office of the Deputy Under Secretary of Defense, Installations and Environment, '2007 Defense Environmental Programs (DEP) Report to Congress' (Washington DC: Department of Defense, 2007), Appendix B, p. 8, available at www.denix.osd.mil/portal/page/portal/denix/environment/ARC.

latter was the initiative whereby the DOD partnered with affected communities and states in the pursuit of 'a broad-based, long-range outreach strategy for range sustainment'.[42] In practice, this meant that the DOD sought to get involved in long-term community planning to ensure the containment of future encroachment issues by, for example, buying up the land of neighbouring communities and creating larger buffer zones; a practice known under the label 'cooperative conservation'. This being said, it is important to note here that the DOD was eager to stress that no forceful relocation of communities would take place, and that this solution was only feasible as long as there were willing sellers.[43] The case of the so-called Northwest Florida Greenway project is perhaps the best practical example of how the DOD's outreach initiative worked. The project is a Memorandum of Partnership (MOP) signed by the DOD, the State of Florida and the Florida Chapter of the Nature Conservancy, aimed 'to conserve environmentally significant lands and limit incompatible development in Northwest Florida'.[44] The project designated a corridor of land and airspace tantamount to Florida's Panhandle region, located between the Apalachicola National Forest and Eglin Air Force Base. The Panhandle region is ecologically valuable in that it contains 75 per cent of Florida's plant species; twenty-three federally endangered and thirteen federally threatened species; old growth longleaf pine forests; and healthy rivers, bays and estuaries. The Eglin Air Force Base, in contributing some $5.9 billion to Florida's economy, is vital for the region's economic development; here, the interest in the preservation of both for the State of Florida is obvious. The MOP between the different parties sought to combine both these needs. Its aims are as follows:

[42] US Department of Defense, *DoD Sustainable Ranges Initiative* (Washington DC: internal publication courtesy of the Office of the Under Secretary of Defense, Installations and Environment, 2005), p. 7.

[43] Donald Rumsfeld, 'Secretary Rumsfeld's remarks at the White House Conference on Cooperative Conservation' (Washington DC: Office of the Assistant Secretary of Defense, Public Affairs, 2005) at www.defenselink.mil/transcripts/transcript.aspx?transcriptid=3119 [1/2006].

[44] Northwest Florida Greenway Memorandum of Partnership among Department of Defense, State of Florida, and the Florida Chapter of the Nature Conservancy to Conserve Environmentally Significant Lands and Limit Incompatible Development in Northwest Florida (2003), p. 2, available at www.floridadep.org/secretary/news/2003/nov/pdf/nwflgw_mop.pdf [10/2005].

- Promote the sustainability of the military mission in northwest Florida to meet national defense testing, operational and training requirements; and
- Protect lands that will sustain the high biodiversity of the region, link protected natural areas, preserve water resources and provide recreation; and
- Strengthen the regional economy by sustaining the mission capabilities of the military in the region and enhancing recreation and tourism[45]

Since signing of the MOP on 12 November 2003, the Northwest Florida Greenway has been very successful, with all parties involved content with its progress. In terms of funding, all parties contribute to the initiative, albeit at different levels. Thus, for example, the initiative was instigated by the State of Florida investing $300 million, whereas the DOD in 2004 set aside $1 million.

Examples such as this one aside, not all was rosy between the DOD and the environmental community in the US since the first Bush administration took office. The main reason for this lay in a combination of some of the remaining pathways to handle encroachment issues and the underlying strategy of exemptions from existing environmental laws. Thus exemptions were what was hoped to be achieved, by pathways such as programming, leadership and organisation, legislation and regulation, and policy. The campaign for exemptions from environmental laws and regulations began in earnest with a plea for certain environmental exemptions in the 2002 annual Defense Authorisation Act. Partly because of a summer 2002 report issued by the General Accounting Office, concerned with the issue of encroachment and the loss of military training, such exemptions were not granted, as the report found that:

Despite the loss of *some* capabilities, service readiness data do *not* indicate the extent to which encroachment has significantly affected reported training readiness. While encroachment workarounds *may* affect costs, the services have *not* documented the overall impact of encroachment on training costs. *Training readiness, as reported in official readiness reports, remains high for most units.* Our analysis of readiness reports from active duty units in fiscal year 2001 showed that very *few* units reported being *unable* to achieve combat-ready status due to inadequate training areas.[46]

---

[45] *Ibid.* p. 2.
[46] United States General Accounting Office, *Military Lacks a Comprehensive Plan*, p. 3 (emphasis added).

Nonetheless, the loss of some capabilities was noted and neither this report nor the 2002 defeat in Congress stopped the DOD from pressing on with its exemption agenda. Indeed, the Sustainable Ranges 2003 Decision Briefing to the Deputy Secretary of Defense, on 10 December 2002, appears to have followed the mantra 'now more than ever'. Under the subheading '2002 lessons learned', some of the bullet points read as follows: long campaign; increase emphasis on regulatory and administrative options; need quantifications of our case; need more 'Operator' participation throughout process; outreach effort needs better definition; variety of views on scope of legislation, and so on.[47] Moreover, the legislative proposals contained in this particular 'Range and Sustainment Package' aimed for the first time ever to:

clarify key aspects of the Migratory Bird Treaty Act (MBTA), the Endangered Species Act (ESA), the Marine Mammal Protection Act (MMPA), the Clean Air Act (CAA), the Resource Conservation and Recovery Act (RCRA), and the Comprehensive Environmental Response, Compensation, and Liability Act (CERCLA) *which have been identified over the past two years as major impediments to training and range sustainment.*[48]

In more detail these 'clarifications of key aspects' meant the following:

MBTA: Exempt military readiness activities from the Migratory Bird Treaty.

ESA: Confirm that Integrated Natural Resource Management Plans can serve in place of critical habitat.

MMPA: Adjust Marine Mammal Act harassment definition to exclude insignificant behavioral change.

CAA: Provide DOD a reasonable time period (5 years) to integrate its emissions with state emission limitations.

RCRA: Clarify that munitions deposited and remaining on operational ranges are not 'solid wastes'.

CERCLA: Clarify that life fire training is not subject to the Comprehensive Environmental Response, Compensation, and Liability Act.[49]

---

[47] Office of the Deputy Secretary of Defense, *Sustainable Ranges 2003 Decision Briefing to the Deputy Secretary of Defense* (Washington DC: Department of Defense, 2002), p. 3.

[48] *Ibid.* p. 5 (emphasis added).

[49] *Ibid.* pp. 17–21.

In other words, the DOD under the first Bush administration sought a mixed strategy of exemptions coupled with the rewriting of long-term environmental laws and regulations that were in place even long before the first Clinton administration took office.

The decision briefing was followed in March 2003 by a memo sent by the Deputy Secretary of Defense Paul Wolfowitz to the Secretaries of the Army, Navy and Air Force respectively. This memo, in tune with the December 2002 briefing, was concerned with environmental exemptions for military training. This time, however, the focus was not so much on seeking new exemptions per se, but rather on making use of existing legislation, which would have allowed the President to exempt the DOD from the above environmental laws, if that was found necessary 'for reasons of national security'.[50]

Throughout 2003 and for some of 2004 DOD's 'campaigning' on the issue of exemptions remained strong. In 2004 their efforts paid off when the Defense Authorization Act granted some of the desired exemptions. This success was facilitated by the fact that the Republicans held a majority in Congress at that time, with the notorious James Inhofe (Republican, Oklahoma) – who had previously called climate change 'the biggest hoax ever perpetrated on the American people'[51] – chairing the Senate Environment and Public Works Committee, whilst known 'friend of the military' Duncan Hunter (Republican, California) chaired the House Armed Services Committee.[52] For the DOD, exemptions could not have come any later than they did; thus by 2004 they were financially stretched to the maximum in the ongoing war effort in Iraq and spending cuts were being liberally suggested across the financial board. The head of the Army's Installations Management Activity command, Major General Anders Aadland, summed up this new thinking in a memo to all garrison commanders in which he recommended savings. With regard to the environment he suggested, 'Take additional risk in environmental programs; terminate environmental

[50] Office of the Deputy Secretary of Defense, 'Memorandum for the Secretary of the Army, the Navy and the Air Force. Subject: Consideration of Requests for Use of Existing Exemptions under Federal Environmental Laws' (Washington DC: Department of Defense, 2003), p. 1, at www.peer.org/docs/dod/Wolfowitz_memo.pdf [10/2005].

[51] James Inhofe, 'The Science of Climate Change Senate Floor Statement', US Senate Committee on Environment and Public Works, 28 July 2003, at http://inhofe.senate.gov/pressreleases/climate.htm [9/2008].

[52] Durant, *The Greening of the US Military*, p. 230.

contracts and delay all non statutory enforcement actions to FY05'.[53] Once the press and watchdogs like PEER got wind of this memo, the statement was revised, and later read: 'Proceed with all previously planned activities within your annual funded program; do not reduce or defer environmental projects'.[54]

In December 2004, marking the last of this type of activity for the first Bush administration, a draft copy of a DOD directive called Environment, Safety, and Occupational Health (ESOH), said to replace the Clinton administrations' Environmental Security Directive, was leaked to the environmental watchdog PEER. In January 2005 the second Bush administration was inaugurated, and in March 2005 the aforementioned new directive became official policy, annulling the DOD Directive 4715.1 Environmental Security for good. Despite this new directive featuring 'environment' in the title, the environment, or rather environmental missions, received little mention in the several pages long document. Indeed, in this new directive, the several paragraphs of the old environmental security directive that outlined the specifics of the role of the DOD for environmental protection in relation to pollution prevention, compliance, conservation and so forth were replaced with just one paragraph: 'It is DOD policy to evaluate all activities for current and emerging ESOH resource requirements and make prudent investments in initiatives that support mission accomplishment, enhance readiness, reduce future funding needs, prevent pollution, prevent illness and injury, ensure cost-effective compliance, and maximize the existing resource capability'.[55]

Given all that has been said here, what can be concluded in terms of the domestic side of environmental security under the Bush administrations? In the case of the Clinton administrations, it was at first difficult to ascertain in what way environmental security was different from prior DOD environmental initiatives. However, the analysis of the scope of the environmental security programme soon revealed that it was different in terms of budget, and that there was a change of

53 Anders Aadland, leaked memo to Garrison staff, 11 May 2004, p. 4 at www.peer.org/docs/dod/armyslashesenvironment.pdf [10/2005].
54 Public Employees for Environmental Responsibility, 'US Army Restores Environmental Funding', 28 May 2004, at www.peer.org/news/news_id.php?row_id=370 [10/2005].
55 Office of the Under Secretary of Defense for Acquisition, Technology, and Logistics, Department of Defense Directive 4715.1E, 'Environment, Safety, and Occupational Health (ESOH)', 19 March 2005, at www.dtic.mil/whs/directives/corres/pdf/471501p.pdf [11/2005].

behaviour regarding leadership and commitment, suggesting a securitisation of the environment as opposed to simply a politicisation. In order to compare and contrast the DOD's initiatives under the Bush administrations with those of its predecessor, it is useful to examine the strategy using the same pointers. This analysis will begin with commitment and leadership, leaving the budget issue till last.

As already shown above, there was generally little commitment to environmental issues on the part of President Bush, and unsurprisingly there was equally little commitment to the environment in the rest of his administrations.[56] Traditionally, Republicans (except Richard Nixon) have always neglected the environment and focused on the economy instead, usually hinting at the incompatibility of the two.[57] With regard to defence environmental security, it could be argued that the same strategy – the claim of incompatibility, if, this time, between national (military) security and environmental rules and regulations – had been invoked. Thus, environmental laws suddenly constituted a risk to, or an obstacle in, the provision of national security (seen again as military security only), because their existence left the armed forces unable to train properly. Indeed, Secretary Rumsfeld's DOD followed the motto: 'We must train like we fight and fight like we train'.[58] This, according to Army Vice Chief of Staff General John Keane,

does not just happen. It requires tough realistic training under demanding battlefield-like conditions to effectively meld soldiers and equipment into the best fighting force in the world. [...] Our soldiers cannot fight with confidence without realistic live-fire and maneuver training. And we need training areas – maneuver land and live-fire ranges – to make this happen. The first time soldiers conduct a realistic operation cannot, cannot, be during time of war. We must train as we intend to fight. And it is becoming increasingly difficult to do so under such environmental restrictions.[59]

[56] On the environmental credibility of members of the Bush administration see for example Pope *et al.*, *Strategic Ignorance*, or Robert S. Devine, *Bush versus the Environment* (New York: Anchor Books, 2004).

[57] Raymond Tatalovich and Mark J. Wattier, 'Opinion Leadership: Elections, Campaign, Agenda Setting, and Environmentalism', in Dennis L. Soden (ed.), *The Environmental Presidency* (New York: State University of New York Press, 1999), pp. 151ff.

[58] Donald H. Rumsfeld, 'Transforming the Military', *Foreign Affairs* 81 (2002), at www.foreignaffairs.com/articles/58020/donald-h-rumsfeld/transforming-the-military [11/2005].

[59] General John Keane, 'The Impact of Environmental Extremism on Military Readiness: The Encroachment Problem' (Washington DC: US Senate Republican

Remembering the history of defence environmental security as outlined in Chapter 3, it comes to mind that environmental damage is often a side effect of war fighting. If, then, training equates to fighting, as Rumsfeld's motto suggests, it appears that such environmental damage had become legitimate. Moreover, the lack of environmental commitment and the incompatibility claim are highlighted by the numerous exemptions the DOD sought under Rumsfeld's leadership. This is so, because at all times – and even without exemptions – it was possible for the President to sidestep environmental rules and regulations if national security had required it. Since the President did not do so, however, it is conceivable to assume that it is *not* the compromise of military training that is at the heart of the Pentagon's campaign, but rather a desire to get rid of environmental laws once and for all. In other words, the desire to get rid of environmental laws may not have originated in the compromised training circumstances alone, but rather in a deep-seated lack of environmental commitment on part of the DOD.

Moving on to leadership, the Bush administrations did not only miss an environmental leader à la Gore, but the ODUSD–I&E missed its environmental leaders too and no longer was the office run by the dedicated Goodman and her deputy Gary D. Vest with his successful, if unconventional, ways. Although the ODUSD–I&E was staffed by many of the same bureaucrats as during the Clinton administrations – many of whom undoubtedly remained committed to environmental stewardship by the military – bureaucrats were tied to the decisions made at the higher level, and regardless of whether they agreed with these decisions. In addition, more decisions (like the sustainable ranges issue) than under the Clinton administration were subject to the Office of the Deputy Secretary of Defense or even the Office of Secretary of Defense as opposed to the heads of the ODUSD–I&E.

Having established both a lack in commitment and in leadership, it is now necessary to take another look at the budget. As already outlined earlier in this chapter, funding in all environmental issue areas – except for conservation – had gone down since the advent of the first Bush administration, in some cases by more than 60 per cent. Furthermore, no stable budget allocation existed, but rather decisions were made each year anew according to need.

Committee, 2003), p. 2, at http://kyl.senate.gov/legis_center/rpc/rpc_040103.pdf [1/2006].

Table 4 *Defense Environmental Programs budget 2001–2009 (in billion US dollars)*[60]

FY 2001	FY 2002	FY 2003	FY 2004	FY 2005	FY 2006	FY 2007	FY 2008	FY 2009 (appropriated)
4.132	3.943	4.247	3.653	3.596	3.874	3.741	4.003	4.250

The DOD explained this decrease in funding by pointing to the successful completion of many programmes. An abstract of the FY 2004 Defense Environmental Programs (DEP) Report to Congress, for example, reads as follows: 'DOD expects that overall funding needs will decrease in future years as defense activities become more sustainable, technological advances make environmental activities more efficient and cost effective, and environmental restoration requirements reach or near completion'.[61] An alternative explanation of the decreasing Environmental Quality budget, however, could be that, given that there were now fewer rules and regulations to comply with, less money to do so was needed. This explanation is backed up by the fact that the largest fall in funding occurred from FY 2003 to FY 2004 – when the total budget fell by nearly $600 million – bearing in mind that 2004 was of course the year when the DOD was granted the long-desired exemptions (see Table 4). Besides, this would also explain why there was no longer a stable long-term budget plan.[62] Hence, why have such budget allocations if there may even be fewer rules to comply with tomorrow?

Naturally Bush's DOD would not have been happy with such accusations. During the years of the first Bush administration the

[60] Data for FY 2001 taken from the Office of Deputy Under Secretary of Defense, Installations and Environment, '2004 Defense Environmental Programs (DEP) Annual Report to Congress', p. 71; data for 2002–2006 taken from the 2005 Defense Environmental Programs (DEP) Annual Report to Congress, p. C-2; data for FY 2007–2009 taken from the 2007 Defense Environmental Programs (DEP) Annual Report to Congress; data for FY 2007–2009 taken from the 2008 Defense Environmental Programs (DEP) Annual Report to Congress. All reports are available at www.denix.osd.mil/portal/page/portal/denix/environment/ARC.

[61] ODUSD–I&E, '2004 Defense Environmental Programs (DEP) Annual Report to Congress', p. 4.

[62] The total budget's relative stability is explained by its inclusion of non-environmental programmes, e.g. REPI, DEIC etc.

DOD's campaign for environmental exemptions caused outrage in the environmental community and the media. The DOD was regularly accused of neglecting their environmental profile and rolling back environmental legislation where they could. Despite such accusations, the Pentagon remained adamant that it was a good steward of the environment and was eager for the press to report more balanced news about its activities. That the power of the press, and by extension public opinion, was taken seriously becomes clear when considering that Raymond DuBois, Bush's first Deputy Under Secretary of Defense for Installations and Environment, wrote an open letter to the broadsheet *USA Today*, after the latter had accused the DOD of environmental neglect.[63] That the press and the public concern for the environment were taken seriously by the Bush administration has been usefully highlighted by Pope *et al.* in the following passage:

Why did Reagan's administration get such a terrible environmental reputation? Because, Bush decided, it had bad public relations: Reagan and his appointees did not sufficiently sugarcoat what they were doing wrong. [...] Bush's rhetoric is far more sophisticated. He regularly has his picture taken in natural settings and wraps his anti-environmental policies in benign names like 'Clear Skies' and 'Healthy Forests'. [Bush's political advisor Frank] Luntz advised Bush and Co. in a leaked memo. '[A]ny discussion of the environment has to be grounded in an effort to reassure the skeptical public that you care about the environment for its own sake – that your intentions are strictly honorable'. So are they all, honorable men – even if Clear Skies results in increased pollution and if Healthy Forests chops down trees to save them from fire [...].[64]

Out of all the changes and exemptions the DOD sought, it was the exemption from the Endangered Species Act (ESA) that caused the biggest outcry in the popular press. Many feared that this would be the end to wildlife conservation on part of the DOD altogether. The Endangered Species Coalition, a national network of hundreds of conservation, scientific, education, religious, sporting, outdoor recreation, business and community organisations, posted the following statement on their website in response to the news:

DOD's proposal would fundamentally change the ESA, eliminating the designation of critical habitat on all lands 'owned or controlled' by the

---

[63] Raymond DuBois, 'Pentagon is a Good Steward of the Environment', *USA Today*, 27 October 2004, p. 12.

[64] Pope *et al.*, *Strategic Ignorance*, pp. 24–5.

military – where some of the best habitat remains for more than 300 species on the brink of extinction. To safeguard this nation's natural heritage – especially the hundreds of imperilled species residing on the 25 million acres of military lands – the DOD must continue to uphold its ESA responsibilities. In addition, the ESA already contains three provisions that provide the DOD with flexibility to work with the Act.[65]

Similarly, Jeff Ruch, Executive Director of the environmental watchdog PEER, which did much work to uncover the Bush administrations' environmental policies especially regarding the role of the DOD, said that: 'This bill [the 2003 Defense Authorization Act] is about as subtle as a Sherman tank. Under this bill, the Pentagon will have immunity to contaminate communities, foul the air and wipe out wildlife'.[66] The DOD was eager to stress that nothing of the sort was the case, and that all they wanted was to move away from ESA's critical habitat requirement towards an Integrated Natural Resources Management Plan (INRMP), a move that would allow them to train in the area specified, whilst at the same time complying with conservation law. Moreover, they stressed that the INRMPs are not administered by the DOD alone, but rather in cooperation with the Fish and Wildlife Service (FWS) located in the Department of the Interior as well as individual states' Fish and Wildlife Services. The following extract outlines the purpose of the INRMP according to the US Fish and Wildlife Service:

INRMPs are based on the principles of ecosystem management. INRMPs provide for the management of natural resources, including fish, wildlife, and plants; allow multipurpose uses of resources; and provide public access where appropriate for those uses, without any net loss in the capability of an installation to support its military mission. To the extent appropriate and applicable for a given installation, an INRMP:

- Integrates conservation measures and military operations
- Reflects cooperation between the FWS, State, and installation relative to the proper management of fish and wildlife resources

---

[65] This statement was downloaded by the author in 2004; it has since been removed from The Endangered Species Coalition's website.
[66] Jeff Ruch, 'Pentagon Files for Environmental Exemptions: Mixed Earth Day Message for Bush Administration' (Washington DC: Public Employees for Environmental Responsibility, 2003).

- Documents requirements for the natural resources budget
- Serves as a principal information source for NEPA documents
- Aids planners and facility managers
- Guides the use and conservation of natural resources on lands and waters under DOD control
- Balances the management of natural resources unique to each installation with mission requirements and other land use activities affecting an installation's natural resources
- Identifies and prioritizes actions required to implement goals and objectives[67]

Although it is undoubtedly true that the DOD would still have had to comply with conservation laws once exempted from the ESA, it is also true that the INRMP and ESA have existed alongside one another since 1997 (when the INRMPs were first created) without problems. According to a confidential source from FWS, the INRMPs are definitely a softer touch on the DOD than critical habitat, as they allow for the military to train without paying attention to endangered species living on military land. More interestingly still, the move towards INRMP and away from ESA also explains the rise in funding in the conservation programme of the Environmental Quality budget. However, the budget increase came once more *not* because of an interest in the environment, but because of the all-important issue of encroachment. Thus, '[t]he anticipated increases in the FY2005 and FY2006 Conservation program funding reflect DOD's new initiative to fund *conservation easements* to prevent encroachment on training bases'.[68]

Given all that has been said here on the issue of domestic environmental security, it emerges that the environmental security policy changed so significantly under the Bush administrations that everything points to a desecuritisation. Before an overall verdict can be reached, however, it is important to analyse what became of international environmental security, particularly because it was here that the most innovative environmental security initiatives took place under Clinton.

---

[67] US Fish and Wildlife Service, 'Integrated Natural Resources Management Plan Fact Sheet' (Washington DC: Department of the Interior, 2004), p. 2.
[68] ODUSD–I&E, '2004 Defense Environmental Programs (DEP) Annual Report to Congress', p. 4.

Environmental security international programmes and initiatives

Department of Defense

Since new goals require new strategies, it is not surprising that the preventative defence strategy of President Clinton's Secretary of Defense Perry became sidelined by a new strategy under Bush, namely 'pre-emptive defence'. The Quadrennial Defense Review of 2001, published some two and a half weeks after 9/11 on 30 September 2001, was the first official document to speak of pre-emption. 'The QDR states that defense of the US homeland is the highest priority for the US military; this was painfully reinforced on September 11th. The US must deter, *preempt*, and defend against aggression targeted at US territory, sovereignty, domestic population, and critical infrastructure, as well as manage the consequences of such aggression and other domestic emergencies'.[69] The 2002 NSS made this shift even clearer. Little mention was made of a preventative strategy, with pre-emption dominating the agenda:

defending the United States, the American people, and our interests at home and abroad by identifying and destroying the threat before it reaches our borders. While the United States will constantly strive to enlist the support of the international community, we will not hesitate to act alone, if necessary, to exercise our right of self-defense by acting pre-emptively against such terrorists, to prevent them from doing harm against our people and our country. [...] The United States has long maintained the option of pre-emptive actions to counter a sufficient threat to our national security. The greater the threat, the greater is the risk of inaction – and the more compelling the case for taking anticipatory action to defend ourselves, even if uncertainty remains as to the time and place of the enemy's attack. To forestall or prevent such hostile acts by our adversaries, the United States will, if necessary, act pre-emptively.[70]

Individual policy-makers endorsed the NSS on various occasions. Bush's first National Security Advisor Condoleezza Rice, for example, argued as follows:

It is [...] true that since 9/11, our Nation is properly focused as never before on preventing attacks against us before they happen. The National Security

---

[69] US Department of Defense, 'Quadrennial Defense Review Report' (Washington DC: Department of Defense, 2001), p. 69 (emphasis added).
[70] *US National Security Strategy*, 2002, pp. 6, 15–16.

Strategy does not overturn five decades of doctrine and jettison either containment or deterrence. These strategic concepts can and will continue to be employed where appropriate. But some threats are so potentially catastrophic – and can arrive with so little warning, by means that are untraceable – that they cannot be contained. Extremists who seem to view suicide as a sacrament are unlikely to ever be deterred. And new technology requires new thinking about when a threat actually becomes 'imminent'. So as a matter of common sense, the United States must be prepared to take action, when necessary, before threats have fully materialized.[71]

And Secretary of Defense Donald Rumsfeld argued as follows:

Defending the United States requires prevention and sometimes pre-emption. It is not possible to defend against every threat, in every place, at every conceivable time. Defending against terrorism and other emerging threats requires that we take the war to the enemy. *The best – and, in some cases the only – defense is a good offence.*[72]

The previous chapter showed that the strategy of preventative defence meant that environmental security was regarded as environmental trust and peace-building in volatile areas. Given that the strategic landscape changed after 9/11 the questions must be: How was international environmental security conceived of in the new circumstances? And, what were its aims? Obviously these questions can only be posited because the international side of the defence environmental pro-grammes continued to exist under the first Bush administration, with the overall programme being called Defense Environmental International Cooperation (DEIC). So-called military-to-military environ-mental programmes continued to exist as well, and funding initially increased significantly under the first Bush administration (see Table 5).

As the domestic experience has shown, however, the bare existence of the same programmes does not necessarily mean that they were the same, or rather that the environment continued to be regarded as a security issue. Indeed, it is plausible to suggest that this was no longer the case. Importantly, under Bush, international environmental cooperation pro-grammes were no longer under the jurisdiction of the environmental

[71] Condoleezza Rice, 'Dr Condoleezza Rice discusses President's National Security Strategy' (Washington DC: Office of the Press Secretary, 1 October 2002), available at http://georgewbush-whitehouse.archives.gov/news/releases/2002/10/20021001-6.html.
[72] Rumsfeld, 'Transforming the Military' (emphasis added).

Table 5 *Defense Environmental International Cooperation budget 2001–2006 (in million US dollars)*[73]

Year	FY 2001	FY 2002	FY 2003	FY 2004	FY 2005 estimate	FY 2006 request
Total	2.4	2.2	2.3	3.3	3.6	3.9

office of the DOD, but rather they became part of the Office of the Secretary of Defense (OSD). This is significant, because it meant that the goal of these programmes was to attend to the Secretary of Defense's aspirations as opposed to those of the ODUSD–I&E. As the 2005 Environment, Safety, and Occupational Health (ESOH) Directive states: 'Support the Security Cooperation Guidance and Strategy of the Secretary of Defense through the Defense Environmental International Cooperation Program and other related international activities, consistent with national security policy'.[74] What the exact goals of the OSD were is not known as these materials are classified. Considering the annual defence reports issued by the ODUSD–I&E up until FY 2004, however, it becomes obvious that these goals were increasingly concerned less with the environment and more with the greater narrative of the war on terror. The 2004 report, for example, sounds little different from other government publications of that time, with the strategic goals of the war on terror looming large, whilst there was little mention of the environment.

The Defense Environmental International Cooperation (DEIC) program is an effective and cost efficient way to share environmental information; counter the proliferation of weapons of mass destruction; partner to maintain access to resources for training and readiness; contribute to interoperability; promote regional cooperation; foster a global military environmental ethic; and improve interagency processes, focus, and integration. In Fiscal Year (FY) 2004, DEIC activities focused on building capacity to mitigate encroachment; preserve training range capabilities; and enhance regional capacity to address natural, accidental, or terrorist caused disasters.[75]

[73] All figures taken from the ODUSD–I&E, '2004 Defense Environmental Programs (DEP) Report to Congress', Section Q, p. Q4.
[74] Department of Defense, 'Environment, Safety, and Occupational Health (ESOH) Directive', 19 March 2005, at www.dtic.mil/whs/directives/corres/pdf/471501p.pdf [5/2005].
[75] ODUSD–I&E, '2004 Defense Environmental Programs (DEP) Annual Report to Congress', Q1.

A rise in funding for DEIC therefore does not automatically mean that increased attention was given to environmental issues, but rather the programme aimed to cater to the wider goals of the war effort, through, for example, maintaining installations and capabilities.[76] This said, it is not clear what exactly has happened to this programme beyond FY 2004. The figures for FY 2005 and FY 2006 in Table 5 are taken from the Defense Environmental Program (DEP) Report to Congress for FY 2004, the last time this programme was mentioned in any annual report to Congress of this type. My own enquiries into the matter at the Office of Deputy Under Secretary of Defense, Installations and Environment did not provide a conclusive answer; highlighting vividly that that office was simply no longer in charge of this programme.

My argument that DEIC programmes, at that time, dealt with the war effort more than with environmental issues is further supported by considering the example of AMEC which was but one of the DEIC programmes. AMEC was still going strong under the Bush administrations and in 2004, with the addition of the United Kingdom, even gained a fourth member. Despite going strong, in 2005 the American director of AMEC Dieter Rudolph argued 'there is less emphasis in the programme on the environment now, than there was before 9/11; today all the emphasis is on installations'.[77] Although it would be tempting to conclude that this change came out of the Bush administrations directly, according to Rudolph one would be misled to do so. Rudolph maintained that AMEC changed because the US partners (the UK, Norway and Russia) would only receive funding with regard to the Global Partnership Programme, which was only concerned with the threat stemming from easily accessible nuclear material (terrorists building dirty bombs) as opposed to the environmental threat caused by nuclear contamination. In the US, so Rudolph, they were only reacting to what the partners did. Nonetheless, the Bush administrations

---

[76] In support of this point see also Rear Admiral John F. Sigler, 'US Military and Environmental Security in the Gulf Region', *Environmental Change and Security Project Report* (Washington DC: The Woodrow Wilson Center, 2005), p. 53, who clearly states that after the attacks of 11 September 2001 the job (or better the rationale for the programme) of those involved in DEIC 'became a little easier' exactly because environmental security became linked to terrorism.

[77] Author telephone interview with American Director of the Arctic Military Environmental Cooperation (AMEC) Dieter Rudolph, 27 October 2005.

clearly had little interest in extending the profile of AMEC by focusing on environmental issues as well. This is telling, because under Clinton AMEC dealt both with the issue of safe storage of spent nuclear fuel and the contamination of the land. In the absence of this second issue it is not surprising that funding for AMEC decreased from formerly the US alone contributing $6 million annually, whilst for FY 2006 the combined sum of all partners was just under $7.5 million; with the US request for 2006 amounting to just $1.4 million.[78] That the Bush administrations had little interest in extending the international profile of environmental security is also supported by the fact that besides the activities of the DOD, no other international environmental security programmes existed.

Department of State

Moving on, let us now turn to the Department of State (DOS). The first thing that needs to be noted when analysing the role of the DOS under the Bush administrations is that the Clinton-created Office of the Under Secretary of State for Global Affairs remained virtually intact under Bush, even beyond his re-election in 2004. Under Bush this office was called the Office of the Under Secretary of State for Democracy and Global Affairs, and for the duration of the two Bush administrations it was headed by Paula Jon Dobriansky. This office had oversight of a variety of global issues, including democracy, human rights, environment, health, population, refugees, women's issues, trafficking in persons, avian influenza and pandemic influenza. Dobriansky was assisted by a group of three assistant secretaries working within the Bureau of Oceans and International Environmental and Scientific Affairs (OES), the Bureau of Democracy, Human Rights and Labor (DRL) and the Bureau of Populations, Refugees, and Migration (PRM) respectively. Although the continuous existence of this office may suggest otherwise, environmental diplomacy did not feature greatly with Bush's DOS. Mention was made of 'diplomatic environmental efforts' with regard to the work of the environmental hubs which continued to exist, but in general OES within DOS

---

[78] ODUSD–I&E, '2004 Defense Environmental Programs (DEP) Annual Report to Congress', Q2.

focused all its energies on the policy of sustainable development.[79] Since sustainable development, like environmental security, has many different interpretations the question is, what did the US government understand by the term? In 2002 Bush's first Secretary of State Colin Powell gave an answer, arguing in a speech that:

When we talk of sustainable development, we are talking about the means to unlock human potential through economic development based on sound economic policy, social development based on investment in health and education, and responsible stewardship of the environment that has been entrusted to our care by a benevolent God.[80]

If we compare Powell's definition with the 'original' (first voiced) definition of sustainable development as found in the 1987 report *Our Common Future* edited by Norwegian Prime Minister Gro Harlem Brundtland, then the environment appears very much secondary in Powell's speech.

Sustainable development is development that meets the needs of the present without compromising the ability of future generations to meet their own needs. It contains *within* it two key concepts: (1) The concept of 'needs', in particular the essential needs of the world's poor, to which overriding priority should be given, and (2) the idea of limitations imposed by the state of technology and social organizations on the environment's ability to meet present and future needs.[81]

Indeed, in light of this definition it seems that the label 'sustainable development' as understood by Powell, has been confused with a different concept altogether, namely 'development'. Powell's speech continued as follows: 'Sustainable development is a compelling moral and humanitarian issue. But sustainable development is also a security

---

[79] Author interview with Claudia A. McMurray, Assistant Secretary, Bureau of Oceans and International Environmental and Scientific Affairs (2005–2008), 8 September 2005, Washington DC.

[80] Colin Powell, 'Making Sustainable Development Work: Governance, Finance and Public–Private Co-operation' (Washington DC: Office of the Secretary of State, 2002), available at http://2001-2009.state.gov/secretary/former/powell/remarks/2002/11822.htm.

[81] Gro Harlem Brundtland (ed.), *Our Common Future: The World Commission on Environment and Development* (Oxford: Oxford University Press, 1987), p. 43 (emphasis added).

imperative. Poverty, destruction of the environment and despair are destroyers of people, of societies, of nations, a cause of instability as an unholy trinity that can destabilize countries and destabilize entire regions'.[82] Interestingly, however, neither the 2002 nor the 2006 NSSs made mention of sustainable development, whilst the term *development* was used some thirty odd times in the respective documents, amongst other usages just in the same way that Powell did in his aforementioned speech.

[T]he United States will use this moment of opportunity to extend the benefits of freedom across the globe. We will actively work to bring the hope of democracy, *development*, free markets, and free trade to every corner of the world. The events of September 11 2001, taught us that weak states, like Afghanistan, can pose as great a danger to our national interests as strong states. Poverty does not make poor people into terrorists and murderers. Yet poverty, weak institutions, and corruption can make weak states vulnerable to terrorist networks and drug cartels within their borders.[83]

Given this, it seems as though the main foreign policy goal regarding global issues was not sustainable development (which would indeed focus on the environment), but rather development – really economic development – which does not necessarily include the environment, indeed can be opposed to environmentalism.

With development being such an important issue, it must be assumed that the level of foreign aid would have gone up considerably since the Clinton administrations. This, however, was not the case. Instead, and as a look at the total budget of USAID, the biggest US foreign aid body, reveals, since the years of the Clinton administrations funding levels have remained much the same, with levels at their lowest since the early 1980s (see Table 6).[84]

In addition, the foreign aid budget is incredibly convoluted, and each administration counts different things under the label foreign aid and funds different institutions and programmes. Carol Lancaster, who served as a deputy administrator of USAID during 1993–1996

---

[82] Powell, 'Making Sustainable Development Work'.
[83] *US National Security Strategy*, 2002, p. 4 (emphasis added).
[84] Jeffrey D. Sachs, 'The Strategic Significance of Global Inequality', *Environmental Change and Security Project Report* (Washington DC: The Woodrow Wilson Center, 2003), p. 33.

Table 6 *USAID budget total 1999–2009 (in billion US dollars)*[85]

Year	FY 1999	FY 2000	FY 2001	FY 2002	FY 2003	FY 2004
Total	8.331	7.738	7.506	7.716	8.763	8.838

Year	FY 2005	FY 2006	FY 2007	FY 2008 (estimate)	FY 2009 (request)
Total	8.971	9.068	7.422	7.527	8.217

and who wrote a book on the foreign aid budget called *Transforming Foreign Aid: United States Assistance in the 21st Century*, argues that sometimes administrations count initiatives and programmes into the foreign aid budget that have nothing to do with foreign aid. The Bush administrations, she said in an interview for this book, was a particular champion of such statistics muddling.[86] Nonetheless, even if the levels of foreign aid under the Bush administrations have not increased significantly and even if their figures for the USAID budget or foreign aid in general were incorrect, they did create a brand new aid scheme, the so-called Millennium Challenge Account (MCA) initiative. This programme began in 2004 'with $1.3 billion in funding and then slated to rise to $5 billion per year by 2006',[87] which, notably, was as much as the environmental security programme received annually under Clinton. What then was the MCA, and what was its purpose? An OES fact sheet on the MCA issued in August 2002 gives the following explanation:

[The] Millennium Challenge Account will fund initiatives to help developing nations that demonstrate a strong commitment to ruling justly, investing in people, and promoting economic freedom, which are the foundations for broad-based, lasting development. [...] The new compact recognizes that economic development assistance can be successful only if it is linked to

---

[85] The data for 1999–2001 is taken from the Summary of USAID Fiscal Year 2001 budget justification; data for FY 2002 from USAID budget justification FY 2002; data for FY 2003 taken from USAID budget justification to Congress from FY 2006 and data for FY 2004–2006 from USAID budget justification to Congress from FY 2007; data for FY 2007–2009 taken from USAID budget justification to Congress FY 2009. All budget justification reports are available at http://www.usaid.gov/policy/budget.

[86] Author interview with Deputy Administrator of the US Agency for International Development (1993–1996) Carol Lancaster, 6 September 2005, Georgetown University, Washington DC.

[87] Sachs, 'The Strategic Significance of Global Inequality', p. 32.

sound policies in developing countries. In sound policy environments, aid attracts private investment by two to one – that is, every dollar of aid attracts two dollars of private capital. In countries where poor public policy dominates, aid can actually harm the very citizens it was meant to help by subsidizing bad policies and delaying reform. The funds into the Millennium Challenge Account will be distributed to developing countries that have demonstrated a strong commitment toward: *Ruling Justly*: Rooting out corruption, upholding human rights, and adherence to the rule of law are essential conditions for successful development. *Investing in People*: Investment in schools and health care provide for healthy and educated citizens who become agents of development. *Promoting Economic Freedom*: More open markets, sustainable budget policies, and strong support for individual initiative will unleash the enterprise and creativity for lasting growth and prosperity. [...] President Bush wants to close the growing divide between nations that are making progress and those that are failing. The MCA will provide significant new support to countries whose governments have made the right policy choices. The President wants to include every man, woman, and child in an ever-expanding circle of development.[88]

In view of this long statement, it appears that the MCA was not in the first instance concerned with magnanimous development goals, but rather it was concerned with failed states and how to prevent more failed states. To prevent, the MCA works on two levels – first, by being essentially a set of rules and, second, by making funding dependent on states operating by those rules. The reasons for this focus can again be found in the greater narrative of the war on terror. In President Bush's words, 'persistent poverty and oppression can lead to hopelessness and despair. And when governments fail to meet the most basic needs of their people, these failed states can become havens for terror'.[89] The MCA and in general the focus on development on the part of the Bush administrations aimed to address these problems, leaving the hierarchy of global issues – including environmental change – to be dictated once more by the war on terror, the grand narrative that had come to dominate everything and that justified most things.

[88] Bureau of Oceans and International Environmental and Scientific Affairs, 'Fact Sheet Millennium Challenge Account' (Washington DC: Department of State, 2002).

[89] George Bush, 'Global Development, President's Remarks to Inter-American Development Bank' (Washington DC: The Office of the Press Secretary, 2002), available at http://georgewbush-whitehouse.archives.gov/news/releases/2002/03/20020314-7.html.

Much later on, towards the end of the second Bush administration, the Secretary of Defense Robert Gates took this narrative one step further, linking terrorism with climate change. '[T]he country is again trying to come to terms with new threats to national security. Rather than one, single entity – the Soviet Union – and one, single animating ideology – communism – we are instead facing challenges from multiple sources: a new, more malignant form of terrorism *inspired* by jihadist extremism, ethnic strife, disease, poverty, climate change, failed and failing states, resurgent powers, and so on'.[90] This move itself, however, was only possible after climate change had become a prominent issue for the second Bush administration;[91] how this happened and what this meant for environmental security will be explained in the next section.

Climate change and energy security

In was not until the summer of 2007 that the environment eventually appeared on the US government's political agenda, when President Bush formally acknowledged the man-made element in global

---

[90] Robert M. Gates, 'Speech to the Association of American Universities' (Washington DC: Office of the Assistant Secretary of Defense, Public Affairs, 2008), available at www.defenselink.mil/speeches/speech.aspx?speechid=1228. Interesting to note about this is that this is not exactly new territory for Gates. In his role as CIA director (1991–1993) he was one of the instigators – alongside Gore – of the MEDEA programme.

[91] In evidence consider the following anecdote noted by myself and Geoffrey D. Dabelko elsewhere: 'In 2003, the Pentagon's long-range planning office headed by Andrew Marshall commissioned risk analyst scenario writers Peter Schwartz and Doug Randall to examine whether climate change may pose a threat to US security. Their dramatic scenario postulated dramatic security implications including endemic "disruption and conflict" from abrupt climate change. Posted on the Pentagon website, the report went little noticed until picked up by *Fortune* and subsequent overseas media coverage that erroneously called the report "secret" and "classified". The resulting public attention to a perceived difference between White House and Department of Defense climate change threat assessments led to the removal of the report from the website and from subsequent Department of Defense comment on climate change': Rita Taureck and Geoffrey D. Dabelko, 'Profile of the United States', in Ronald A. Kingham (ed.), *Inventory of Environment and Security Policies and Practices: An Overview of Strategies and Initiatives of Selected Governments, International Organisations and Inter-Governmental Organisations* (The Hague: Institute for Environmental Security, 2006), p. 3.

warming.[92] Dr Sharon Hays, Associate Director of the White House's Office of Science and Technology Policy, explained Bush's changed position on this issue as follows:

[T]here's a lot of misconceptions out there about the administration's position on this. The President gave a speech back in 2001, I believe it was, in which he articulated his thoughts on climate change science. And when he made that speech, he echoed very clearly what the National Academies of Science was saying about climate change and the human attribution issue, which was, at that time, six years ago, less clear than it is now. And so his words echoed the National Academies of Science, which, in turn, were really a restatement of what appeared in the last assessment, the 2001 IPCC assessment on climate change. So the President has been saying the same thing as what the scientists have been saying about climate change, whether or not humans are playing a role in it, since 2001. More recently, the President made a speech at the Major Economies meeting that he convened where, again, he cited the IPCC directly in articulating the administration's view on climate change science. So it is and has been absolutely in sync with what the science is saying.[93]

As will be shown, Bush's newfound position on climate change, however, did not translate into actual care for the environment. It was

[92] This was a major shift in the rhetoric of the Bush administration. Thus when in 2002 the EPA submitted a report to the United Nations that acknowledged that man-made pollution is largely to blame for global warming, this report was dismissed by President Bush a few days later. In the words of former EPA climate policy adviser in the Office of Air and Radiation Jeremy Symons: 'White House officials tried to force the EPA to alter the scientific content of the report in order to play down the risks of global warming. The EPA has billed the report, released in June, as "the first-ever national picture of environmental quality and human health in the United States." An internal EPA decision paper noted that White House officials were insisting on "major edits" to the climate change section and were telling the EPA that "no further changes may be made" beyond the White House edits. In the internal paper, EPA staff warned that the report "no longer accurately represents scientific consensus on climate change". The EPA ultimately pulled the global warming section from the report to avoid publishing information that is not scientifically credible': Jeremy Symons, 'How Bush and Co. Obscure the Science', *Washington Post*, Sunday, 13 July 2003; see also Lloyd de Vries, 'Bush Disses Global Warming Report', *CBS News*, 4 June 2002, and Paul Harris, 'Bush Covers up Climate Research', *The Observer*, 21 September 2003.

[93] James Connaughton, Sharon Hays and Harlan Watson, 'Press Briefing Via Conference Call by Senior Administration Officials on IPCC Report' (Washington DC: Office of the Press Secretary, 2007) available at http://georgewbush-whitehouse.archives.gov/news/releases/2007/11/20071116-21.html.

rather the case that the second Bush administration had realised that there was a vast gap between their take on climate change and that of the majority of the American people, not to mention the international community. In the US many states (including Republican-governed California), as well as major industries, were opposed to Bush's suggestion of a voluntary reduction in carbon dioxide emissions. California, for example, has set a fixed target of a 25 per cent reduction in carbon emissions by 2020. A 2006 poll by the Public Policy Institute of California shows that this move was clearly supported by Californians.[94] In response to these trends the second Bush administration took great care to portray themselves as champions of climate action. Indeed they purposely set out to sell their real aim – energy independence – as a concern for climate change. In the wider context of this book it is important to note that the second Bush administration's decision to take climate change on board *after* sensing public discontent over its lack of concern for this issue confirms my earlier reconfiguration (pp. 49–52) of the audience with the electorate. With climate change nearer the forefront of the American consciousness than most other environmental issues past and present, and with elections due in November 2008, the second Bush administration could simply not afford to ignore this issue. Climate change subsequently featured prominently in the 2008 presidential election campaigns.[95]

Concern for energy independence had come to the forefront only during the second Bush administration. Mentioned only in passing in the 2002 NSS, energy security and independence from foreign oil were major concerns in the 2006 NSS. There were two reasons for this. First, the ratification of the 2005 Energy Policy Act had restructured America's energy policy, offering 'consumers and businesses federal tax credits beginning in January 2006 for purchasing fuel-efficient hybrid-electric vehicles and energy-efficient appliances and products'.[96] Second, the war in Iraq and the larger US government goals for spreading democracy in the Middle East had raised concern about

---

[94] Felicity Barringer, 'Officials Reach California Deal to Cut Emissions', *New York Times*, 31 August 2006.

[95] Elisabeth Bomberg and Betsy Super, 'The 2008 US Presidential Election: Obama and the Environment', *Environmental Politics* 18 (2009), pp. 424–30.

[96] US Department of Energy, 'The Energy Policy Act of 2005 Tax Credits' (Washington DC: Department of Energy, 2005) at www.cepc.net/EnergyPolicyActTaxCredits2005.pdf [6/2007].

US dependence on foreign sources of oil and what those petrodollars were funding. President Bush's 2006 State of the Union address high-lighted this concern well: 'Keeping America competitive requires affordable energy. And here we have a serious problem: America is addicted to oil, which is often imported from unstable parts of the world. The best way to break this addiction is through technology'.[97] By January 2007 the narrative surrounding energy security had become fully established and the issue dominated that year's State of the Union address. 2007 was of course the year that the second Bush administration accepted the science behind climate change, and from then on in energy security and climate change were always bracketed together.

In an effort to achieve energy security the second Bush adminis-tration proposed a number of initiatives and policies, yet a closer look at these initiatives reveals that they did very little if anything to actually curb greenhouse gases. For instance, in December 2007 Bush signed the Energy Independence and Security Act (EISA), which responded to his 'Twenty in Ten' challenge in the 2007 State of the Union address to increase the use of biofuels.[98] From an environ-mental (security) perspective, however, the use of biofuels is highly problematic. A number of independent studies suggest that there is a clear link between increased demands for biofuels in the Global North and food insecurity in the Global South. Coupled with social effects these food shortages are believed to increase the likelihood of environ-mentally induced conflict.[99] In addition, studies conducted by UN Energy and the Swiss Government show that the production of bio-fuels can have serious negative environmental consequences, including loss of biodiversity and increased ground acidity. Worst of all, the production of biofuels is believed to counterbalance greenhouse gas production. This counterbalancing can be either direct in that 'changes

---

[97] President George W. Bush, 'State of the Union', 31 January 2006, at http://georgewbush-whitehouse.archives.gov/stateoftheunion/2006/ [2/2006].

[98] President George W. Bush, 'State of the Union', 23 January 2007, at http://georgewbush-whitehouse.archives.gov/news/releases/2007/01/20070123-2.html [1/2007].

[99] Oxfam Briefing Note, 'Bio-fuelling Poverty: Why the EU Renewable-fuel Target may be Disastrous for Poor People' (Oxford: Oxfam International, 2007); Friends of the Earth Netherlands, Lembaga Gemawan Indonesia and KONTAK Rakyat Borneo, 'Policy, Practice, Pride and Prejudice' (Amsterdam: Friends of the Earth Netherlands, 2007).

in the carbon contents of soils, or carbon stocks in forests and peat lands related to bioenergy production, might offset some or all of the GHG benefits',[100] or indirect, for example through the clearing of healthy forests to make way for biofuel plantations.[101]

Another initiative championed by the second Bush administration has been the greatly advertised Asia–Pacific Partnership on Clean Development and Climate (APP).[102] This initiative has focused on 'voluntary practical measures [...] to create new investment opportunities, build local capacity, and remove barriers to the introduction of clean, more efficient technologies'.[103] Notably, climate change per se has not been the most important element in this programme. Rather, and as the participating governments were eager to stress, the APP does not only focus on climate change, but proposes an integrative approach that views access to energy resources (energy security), poverty alleviation, sustainable development and global climate change as parts of a larger integrated approach. According to Bush's Under Secretary of State for Democracy and Global Affairs Dobriansky, APP is about integration, collaboration and implementation.[104] Critics, however, doubt the effectiveness of the APP. An April 2006 progress report by the Climate Institute of Australia, for example, has warned that 'even in the most optimistic scenarios the agreement would lead to a doubling of greenhouse gas emissions from present levels by 2050, rather than the drastic reductions that are needed to quell climate change'.[105]

---

[100] UN Energy, 'Sustainable Bio-energy: A Framework for Decision Makers' (New York: United Nations, 2007), p. 43.

[101] Friends of the Earth Netherlands *et al.*, 'Policy, Practice, Pride and Prejudice'.

[102] This initiative is also known as AP6, the '6' referring to the six countries that make up this partnership. These are, besides the US, Australia, Japan, India, China and the Republic of Korea.

[103] George W. Bush, 'President Bush and the Asia-Pacific Partnership on Clean Development' (Washington DC: Office of the Press Secretary, 2005) available at http://2001-2009.state.gov/g/oes/rls/fs/50314.htm.

[104] Paula Dobriansky and James Connaughton, 'Briefing: US Participation in the Asia-Pacific Partnership on Clean Development and Climate Change' (Washington DC: Department of State, 2006), available at http://2001-2009. state.gov/g/rls/rm/58780.htm.

[105] Cited in Gabrielle Walker and Sir David King, *The Hot Topic: How to Tackle Climate Change and Still Keep the Lights On* (London: Bloomsbury, 2008), p. 213.

There were other Bush administration international climate change initiatives. For example, the International Partnership for a Hydrogen Economy aimed at providing 'a mechanism to organise, evaluate and coordinate multinational research, development and deployment programs that advance the transition to a global hydrogen economy'.[106] The Carbon Sequestration Leadership Forum (CSLF) focused on 'the development of improved cost-effective technologies for the separation and capture of carbon dioxide for its transport and long-term storage'.[107] The US government led the Generation IV International Forum, concerned with research into 'the next generation of safer, more affordable, and more proliferation-resistant nuclear energy systems. This new generation of nuclear power plants could produce electricity and hydrogen with substantially less waste and without emitting any air pollutants or greenhouse-gas emissions'.[108] Finally, under the 2006 'Advanced Energy Initiative', the DOE was to receive a 22 per cent increase – a total of $103 million – in funding for alternative energy sources, aimed at reducing fossil fuel dependence by 75 per cent by the year 2025.[109] That the second Bush administration's commitment to climate change was rather weaker than these initiatives suggest was apparent at the US-initiated first 'Major Economies Meeting on Energy Security and Climate Change' in September 2007, where binding emission targets were not an issue despite Bush's claims that 'Energy security and climate change are two of the great challenges of our time. [And that] the United States takes these challenges seriously'.[110] In the aftermath of the meeting the *Guardian* newspaper reported one European diplomat as saying: 'It was a total charade and has been exposed as a charade. [...] I have never heard a more humiliating speech by a major leader. He [Mr Bush] was trying to present himself as a leader while

---

[106] George W. Bush, 'Climate Change Fact Sheet' (Washington DC: Office of the Press Secretary, 2005), http://georgewbush-whitehouse.archives.gov/news/releases/2005/05/20050518-4.html.

[107] *Ibid.*

[108] *Ibid.*

[109] Julian Borger, 'Bush Hits the Road to Take a Green Message to his Nation of Oil Addicts', *The Guardian*, 2 February 2006.

[110] George W. Bush, 'President Bush Participates in Major Economies Meeting on Energy Security and Climate Change' (Washington DC: Department of State, 2007), available at http://2001-2009.state.gov/g/oes/rls/rm/2007/92938.htm.

showing no sign of leadership. It was a total failure'.[111] Overall that
meeting was widely regarded as a US effort to spoil the upcoming UN
conference on climate change in Bali in December that year.[112] In Bali
itself, the US delegation – together with Canada and Japan – was
reluctant to sign up to European (and other countries') proposed binding targets of a 25 to 40 per cent carbon emissions reduction, compared
to 1990 levels, by 2020 and, partly as a result of this, the so-called 'Bali
Roadmap' does not feature any binding targets. Under heavy criticism
from other countries and high-profile individuals (notably newly
adorned Nobel Peace Prize laureate Gore) the US delegation maintained throughout the negotiations that Bali was never considered a
forum to discuss binding targets, but rather 'the first step before you can
then sit down and work through the specifics of what that goal might
be'.[113] This kind of evasiveness was typical of the US delegation's
handling of questions regarding binding targets, with all three senior
negotiators (Senior Climate Negotiator and Special Representative at
the US Department of State Dr Harlan Watson; the Chairman of the
White House Council on Environmental Quality James Connaughton;
and the Under Secretary of State for Democracy and Global Affairs
Paula Dobriansky) dubiously arguing that outcomes should not be
prejudiced from the outset by setting targets. The remainder of these
Bali press conferences was spent with the three staff outlining the
virtues of the Bush administrations' various energy and climate action
programmes, denying that there was disagreement between the US and
other countries on the importance of climate change and at all times
emphasising that since 2001 the Bush administrations invested some
$37 billion into climate change related activities. The total allocation of
federal funds on this issue for FY 2008 amounted to $7.37 billion.

These figures notwithstanding, it is important to realise that the Federal
Budget did not actually include a category for spending on climate change
per se. Rather, since 2006 the Office of Management and Budget (OMB)
has produced a 'Federal Climate Change and Expenditure Report to

111 Anonymous, cited in Ewan MacAskill, 'Europeans Angry after Bush Climate
Speech Charade', *The Guardian*, 29 September 2007.
112 See, for example, Julian Borger, David Adam and Suzanne Goldenberger, 'Bush
Kills off Hopes for G8 Climate Change Plan', *The Guardian*, 1 June 2007.
113 Paula Dobriansky, James Connaughton and Harlan Watson, 'December 13
Press Conference by the US delegation' (Bali: Department of State, 2007),
available at http://2001-2009.state.gov/g/oes/rls/rm/2007/97472.htm.

Congress' that accounts for all spending on climate change related activities. Miriam Pemberton from the Institute for Policy Studies has analysed federal spending on climate change in some detail. She argues:

A couple of caveats about this figure: the OMB report correctly notes the difficulty of specifying exactly what the federal government spends to address climate change, acknowledging that some expenditures 'are not solely for climate change purposes', although 'they can provide climate change benefits'. While there are undoubtedly judgment calls to be made about what programs to include in a list of federal climate change expenditures, it's clear that some items in the government's list don't belong. One wonders, for example, what a $1.8 million program to eradicate illegal coca in Peru has to do with climate change. It's also the case that the 2006 OMB climate change report – the first – provides a greater level of detail about the programs included than the 2007 report. In other words, between the first and second reports, reporting became less transparent, and unjustified expenditures easier to hide.[114]

Pemberton further reveals that for FY 2008 $1.83 billion was spent on research into 'the reduction of fundamental scientific uncertainties associated with climate change'.[115] That is, a quarter of the budget was spent on verifying that climate change actually takes place. The largest chunk of the budget, $3.917 billion, was spent on technology development across seven different agencies; this included the government's initiatives discussed above. Above all it is important to realise that this budget did not include any binding targets, leaving Pemberton to conclude that, 'in combination with a policy of opposing such federally mandated ceilings, focusing on the uncertainties of climate science, and opposing changes to tax incentives that will favour renewable sources of energy over fossil fuels, the Bush administration's concentration of the climate change budget on technological development becomes one more diversionary, *delaying tactic*'.[116]

Whatever the case may have been, on 16 April 2008 Bush then actually announced a mandatory emissions target 'to stop the growth of US greenhouse gas emissions by 2025'.[117] In July that year the G8

[114] Miriam Pemberton, 'The Budgets Compared: Military vs. Climate Security' (Washington DC: Institute for Policy Studies, 2008), pp. 19ff.
[115] *Ibid.* p. 24.
[116] *Ibid.* p. 24 (emphasis added).
[117] George W. Bush, 'President Bush Discusses Climate Change' (Washington DC: Office of the Press Secretary, 2008), available at http://georgewbush-whitehouse.archives.gov/news/releases/2008/04/20080416-6.html.

summit declaration on Environment and Climate Change stated a goal of achieving at least a 50 per cent reduction of global emissions by 2050. Yet it was not clear whether the latter referred to 1990s levels or today's emissions levels which are already 25 per cent higher. Moreover, the White House's website issued the latter statement complete with a fact sheet on 'Taking additional action to confront climate change', reminding us, that '[t]here is a right way and a wrong way to approach reducing greenhouse gas emissions [...] the wrong way is to [...] demand sudden drastic emissions cuts that have no chance of being realized and every chance of hurting the economy'.[118] In any case, regardless of the nature of these targets, they came so late in the life of the second Bush administration that *New York Times* foreign affairs columnist Thomas Friedman described George Bush as the 'lamest of lame ducks, he is barely quacking any more'.[119] Putting any kind of targets into action was quite clearly left to the next administration.

This said, Bush was notably far from being a 'lame duck' when it came to energy plans. In a speech on 18 June 2008 three of his four key points to ensure energy security were: first, offshore drilling for oil and gas in the Outer Continental Shelf – a wildlife rich area protected by congressional moratorium since the 1980s; second, further expansion of American oil production by permitting exploration in the Arctic National Wildlife Refuge; and third, petroleum production from oil shale.[120] US oil shale deposits are vast, recoverable resources amounting to more than triple the proven oil reserves of Saudi Arabia; at current consumption rates this equates to more than 400 years of energy security. However, petroleum production from oil shale brings with it a large list of critical environmental impacts, including local air and water pollution and critical habitat loss. Besides, in the words of an official report commissioned by the DOE on petroleum production from oil shale in the United States, 'the production of petroleum

---

118 George W. Bush, 'Fact Sheet: Taking Additional Action to Confront Climate Change' (Washington DC: Office of the Press Secretary, 2008), available at http://georgewbush-whitehouse.archives.gov/news/releases/2008/04/20080416-7.html.

119 Interview with Thomas Friedman, BBC2 *Newsnight*, 17 April 2008.

120 George W. Bush, 'President Bush Discusses Energy' (Washington DC: Office of the Press Secretary, 2008), available at http://georgewbush-whitehouse.archives.gov/news/releases/2008/06/20080618.html.

products derived from oil shale will entail significantly higher emissions of carbon dioxide, compared with conventional crude oil production and refining. In addition, the high temperatures associated with surface retorting can cause a release of carbon dioxide from mineral carbonates contained in oil shale'.[121] Given these and other serious concerns, notably its questionable economic viability, the report concludes – quite contrary to Bush – 'that the future of oil shale remains *uncertain*'.[122]

Despite that, the President was not the only person to promote petroleum production from oil shale as a solution to energy dependence and indeed to global climate change – the Department of Defense did and still does likewise. The Office of the Under Secretary of Defense (Advanced Systems and Concepts) together with the Office of the Under Secretary of Defense (Acquisition, Technology and Logistics) pursues an aggressive strategy to 'catalyze the industry to produce fuels for the military from domestic energy resources (up to 300,000 barrels per day)' much of which is to come from oil shale.[123] The OSD's Assured Fuel Initiative estimates that 2.3+ trillion barrels of oil can be recovered from domestic sources, 1.4 trillion of these from oil shale, dubbing oil shale rich regions in the US the 'New Middle East'.[124] In a written testimony to the Senate's Committee on Energy and Natural Resources in April 2005 Assistant Deputy Under Secretary of Defense (Advanced Systems and Concepts) Theodore Barna explained: 'In 2004 I expanded this initial effort into a wider variety of resources for the production of *clean fuels*, notably oil shale, coal, biomass and petroleum coke by looking at the broader picture of alternative fuels and established the Clean Fuel Initiative'.[125]

[121] James T. Bartis, Tom LaTourette, Lloyd Dixon, D. J. Peterson and Gary Cecchine, *Oil Shale Development in the United States: Prospects and Policy Issues*, RAND Infrastructure, Safety, and Environment Report Prepared for National Energy Technology Laboratory of the US Department of Energy (2005), p. 40.

[122] *Ibid.* p. 53 (emphasis added).

[123] Office of the Secretary of Defense, 'Assured Fuels Initiative Slide Show' (Washington DC: Department of Defense, 2006), p. 1 at www.trbav030.org/pdf2006/265_Harrison.pdf [6/2008].

[124] *Ibid.* pp. 6, 20.

[125] Theodore Barna, 'Written Testimony on Oil Shale and Oil Sands Resources Hearing', Senate Committee on Energy and Natural Resources, 12 April 2005 at http://energy.senate.gov/public/index.cfm?FuseAction=Hearings.Testimony&Hearing_ID=e26d41f4-2d24-4d5a-a5d0-b25e6fef23c1&

Considering the findings of the above mentioned DOE-commissioned report on petroleum production from oil shale (which is backed by similar studies elsewhere[126]), it is not clear how oil shale can qualify as a clean fuel.

Nevertheless, as part of a series of so-called 'midnight regulations' President Bush, in November 2008, opened up some 800,000 hectares of land in Rocky Mountain states for the development of oil shale. This law came into effect on 17 January 2009, three days before Barack Obama took office. To many environmentalists these regulations are the final nail in Bush's environmental coffin. *The Guardian*'s George Monbiot, for example, writes: 'His [Bush's] midnight regulations, opening America's wilderness to logging and mining, trashing pollution controls, tearing up conservation laws, will do almost as much damage in the last 60 days of his presidency as he achieved in the foregoing 3,000'.[127]

The latest statements and actions by the President and DOD on energy security were further confirmation of the Bush administrations' lack of commitment to tackling climate change and to safeguarding the natural non-human environment more generally. Indeed, in these plans energy security and climate (environmental) security were diametrically opposed goals.

How did they desecuritise?

Considering all that has been said here, it can now be concluded that the Bush administrations did not consider the environment a security issue. Further testimony to this is the fact that the many environmental disasters on US territory in 2005 – a year, according to the *New Scientist*, characterised by its many natural disasters – were not cast in environmental security terms. Instead, President Bush in the aftermath of Hurricane Katrina, responsible for the partial destruction of

Witness_ID=e9590f47-9561-49cc-9a61-58b96c700bdb [6/2008] (emphasis added).
[126] See, for example, European Academies Science Advisory Council, 'A Study on the EU Oil Shale Industry: Viewed in the Light of the Estonian Experience', a report by EASAC to the Committee on Industry, Research and Energy of the European Parliament, 2007.
[127] George Monbiot, 'The Planet is Now so Vandalized that only Total Energy Renewal Can Save Us', *The Guardian*, 25 November 2008.

New Orleans and the displacement of thousands of people, linked up this natural disaster with the all-encompassing narrative of the war on terror. The following statement is taken from a White House fact sheet issued on 15 September 2005, some two and a half weeks after the hurricane hit the US Gulf Coast.

The President has ordered The Department of Homeland Security to conduct an immediate review of preparedness in every major American city. Our cities must have clear and up-to-date plans for responding to natural disasters, disease outbreaks, or terrorist attack. We must have plans to evacuate large numbers of people in an emergency and to provide food, water, and security as needed. In a time of terror threats and weapons of mass destruction, the danger is greater than a fault line or flood plain. Emergency planning is a national security priority.[128]

The existence of, and the priority for responding to, natural and man-made environmental disasters was reflected in the revised NSS issued in late March 2006, which, unlike its 2002 predecessor, made reference 'to environmental destruction, whether caused by human behaviour or cataclysmic mega disasters such as flood, hurricanes, earthquakes or tsunamis'.[129] However, whilst here environmental issues returned to the NSS, it is important to note that they did so, not as issues of environmental security, but rather as issues of *general emergency*. This is to say natural disasters mirror the result of a large-scale terrorist attack, for which the country needs to be prepared. This sentiment was echoed by the 2006 Quadrennial Defense Review that recounts in some detail how the DOD supported the Department of Homeland Security in natural disaster relief at home after hurricanes Katrina and Rita, and abroad after 2004 tsunami and the 2005 earthquake in Pakistan.

At the direction of the President or Secretary of Defense, the Department supports civil authorities for designated law enforcement and/or other activities and as part of a comprehensive national response to prevent and

---

[128] George W. Bush, 'Fact Sheet: President Bush Addresses the Nation on Recovery from Katrina' (Washington DC: Office of the Press Secretary, 2005), available at http://georgewbush-whitehouse.archives.gov/news/releases/2005/09/20050915-7.html.

[129] 'The National Security Strategy of the United States of America', March 2006, p. 47, available at http://georgewbush-whitehouse.archives.gov/nsc/nss/2006.

protect against terrorist incidents or to recover from an attack or a disaster. As discussed, the Department's substantial humanitarian contributions to relief efforts in the aftermaths of Hurricanes Katrina and Rita fall into this category. In the future, should other catastrophes overwhelm civilian capacity, the Department may be called upon to respond rapidly with additional resources as part of an overall US Government effort. In order to respond effectively to future catastrophic events, the Department will provide US NORTHCOM with authority to stage forces and equipment domestically prior to potential incidents when possible. The Department will also seek to eliminate current legislative ceilings on pre-event spending.[130]

Overall the various approaches to environmental security that were rhetorically acknowledged by the Clinton administrations all but vanished under the Bush administrations and the verbal connection between 'security' and 'environment' was no longer made. Such a move, whereby a formerly securitised issue is removed from its reference to existential threat, can be called a 'desecuritising move'. Unlike its opposite the securitising move, however, a desecuritising move is by definition a silent process. A desecuritising move is accompanied by the disappearance of security practice. Then, according to the analytical framework here developed, the issue either becomes a political concern for the administration in power and results in politicisation, or whatever was done in the name of the security policy leaves the administration's political agenda altogether – depoliticisation. It remains to analyse what form the Bush administrations' desecuritisation of the environment took.

With regard to the international environmental security policies and practices, a straightforward case for depoliticisation is easily made. Hence the programmes once championed under that label had either disappeared altogether (e.g. the MOU) or they had been transformed to such an extent that they no longer primarily dealt with the environment but with the war on terrorism instead (e.g. DEIC and AMEC). Gore's intelligence environmental security initiative had largely been disbanded. However, this programme has achieved a very low level of institutionalisation in terms of environmental data gathering and the CIA continues to collect data on environmental issues such as forest reductions and droughts. This said, my email requests to the CIA to talk with someone who dealt with the environmental data collection at the

---

[130] US Department of Defense, 'Quadrennial Defense Review Report' (Washington DC: Department of Defense, 2006), p. 26.

time of the Bush administrations were declined because 'everyone was so busy, working on the war on terror'. Furthermore the vast majority of the Bush administrations' foreign aid budget was committed to help with the war on terror. Plus, the work of the Political Instability Task Force (comprised of largely the same group of people that made up the State Failure Task Force Group under Clinton) no longer incorporated the environment as part of their analysis; the focus now was on terrorism.

Like all branches of government the Department of State too became consumed by the war on terror. The analysis showed that although there was much talk of 'sustainable development' in foreign policy circles, suggesting that the environment continued to be a concern for the Bush administrations, their understanding of sustainable development had very little to do with the environment and is more aptly captured by the phrase 'economic development'. The analysis further showed, however, that the administrations were concerned with the latter only in so far as underdevelopment of this kind may breed terrorism.

The war on terror also informed the Bush administrations' handling of perhaps the biggest environmental crisis of all time, global climate change. After many years of openly doubting the science behind man-made global warming, 2007 saw a turn-around in the Bush administrations' discourse on climate change when they finally acknowledged that link. My investigation into the second Bush administration's climate change policies quickly revealed, however, that their sole concern was with energy security. In particular Bush sought to gain independence from foreign oil from volatile regions such as the Middle East. To achieve this aim the second Bush administration championed a number of technological programmes all aiming at energy independence. In light of the international and domestic interest in climate change, however, these policies were cleverly disguised as initiatives to combat climate change, therefore as environmental policies, while binding carbon emission targets – short of some yet to be invented geo-engineering fix as probably the world's only chance to tackle global climate change – were not considered. In addition, as part of his energy security plans Bush championed a number of policies (for example, biofuels and petroleum production from oil shale) which will increase global greenhouse gas emissions, not to mention other localised environmental damage.

With regard to domestic environmental security programmes, the question whether or not environmental stewardship by the military has become politicised after the desecuritisation of environmental security is a little trickier to answer. Thus, although – and as this chapter shows – the case for desecuritisation of defence environmental security is straightforward, it is more difficult to establish whether this has also led to a depoliticisation of defence environmental issues. This is particularly so because there still were defence environmental programmes and there still was money set aside for these programmes. Although these two issues may at first indicate that the issue has become politicised, one should bear in mind that much of what was part of the military's environmental programmes – conservation, compliance and cleanup – was in existence long *before* environmental security ever was an issue. This in turn may suggest two things. First, the only time environmental issues were politicised was when they first came into being (in any case before the Clinton administrations securitised these issues); and second, nowadays they are simply part of existing legislation, i.e. not securitised, not politicised, but simply institutionalised. Indeed, going further than this it seems that the only time environmental issues were politicised on the part of Bush's DOD was when environmental legislation was in the verbal 'firing zone' of those who saw environmental legislation as an obstacle to military training, and therefore to the provision of national security in the war on terrorism.[131] By definition then, it was not the environment that was politicised, let alone securitised, under the Bush administrations; rather, the issue of concern was military readiness. The Bush administrations did nothing to foster environmental legislation; indeed they ditched environmental legislation even if it proved no disadvantage to the military at all, simply, it seems, because they did not approve of it. A politicisation of environmental issues would look different altogether, and it is feasible to conclude that in the domestic realm formerly environmental security practices too have become depoliticised. As they had been during the Cold War, so under the war on terror, military security and defence environmental security once again became interlocked in a zero sum game, with an easily identifiable winner.

Conclusion

This chapter has analysed what happened to the Clinton administrations' environmental security policies and practices under the Bush

---

[131] Compare with Durant, *The Greening of the US Military*, p. 236.

administrations from 2001 to 2009. It argues that the environment not only became desecuritised but that the issue became depoliticised; meaning that the policies and practices that formerly made up environmental security *vanished* from the two Bush administrations' political agendas. Whilst the rise of environmental security under the first Clinton administration could not be understood without the end of the Cold War, the disappearance of environmental security cannot be understood without the context of the events of 9/11 and the subsequent 'war on terror'. What US national security policy may have looked like without these events we cannot know for certain, although, considering the first Bush administration's low estimate of environmental issues, environmental security may have been destined to vanish anyway. Be that as it may, the events of 9/11 closed national security policies to minority discourses of danger such as environmental security. The ensuing war on terror brought with it plenty of funding and new missions for the security establishment. Indeed, as Richard Jackson points out: 'In the "war on terrorism", all of the national security institutions – the military, law-enforcement agencies and intelligence organisations – have received massive extra funding directly because of the fear of terrorist attacks. In America, more than half of the federal budget for FY2004 was devoted to national defence, with the Pentagon receiving $399 billion [...]'.[132]

Finally, let us consider what all this means for securitisation theory and for the theoretical purposes of this book. Recall first of all that original securitisation theory holds that desecuritisation always leads to politicisation, and that my revised securitisation theory proposes a redefinition – a narrowing – of the definition of politicisation to rest with official political authority only. This in turn makes it possible to suggest that desecuritisation can lead to politicisation as well as to depoliticisation, where the latter is understood as the issue vanishing from the political agenda of those with official political authority. The case of environmental desecuritisation under the Bush administrations confirms this logic, as in this case desecuritisation resulted in depoliticisation. Considering, however, that I work with a different definition of politicisation than the Copenhagen School, this case study does not in any strict sense offer a counter-example to the Copenhagen School's

---

[132] Richard Jackson, *Writing the War on Terrorism: Language, Politics and Counter-Terrorism* (Manchester: Manchester University Press, 2005), p. 116.

analysis. Indeed the size of the US government alone means that had the Copenhagen School, or someone informed by their theory, conducted this same analysis they might have found evidence of politicisation. Thus within such a vast structure of government as is the US government there are always some individuals left that continue to work on some of the same issues behind the scenes and under different names.[133] For example, May 2007 saw the creation of a Deputy Directorate for Energy and Environmental Security in the Office of Intelligence and Counterintelligence within the DOE, headed by the former CIA careerist Carole Dumaine. Despite the fact that this office reintroduced the label 'environmental security' into the US government, it is important to note that the office was not a White House creation. Instead it is the sole initiative of the new Director for Intelligence and Counterintelligence Rolf Mowatt-Larssen, who developed a strong personal interest in these issues. In an interview for this book, Dumaine herself described her office as 'a small-scale basement operation, which – when I first started – had neither staff nor a budget', adding that her office 'does not portray or indeed affect the general mission of DOE, let alone, that of the wider government'.[134] In short, the existence of this new position did not mean a re-politicisation of environmental security issues in the US government.

Given the different definition of politicisation I propose, the question we need to ask is, what is the added value for security analysis of narrowing the definition of politicisation in the way here suggested? The short answer is that it leaves the security analyst in a much better

---

[133] The Woodrow Wilson Center's Geoffrey Dabelko, for example, has recently argued that 'the antipathy of US President George W. Bush's administration to anything dubbed "environmental" set back efforts in international forums and pushed much of official US work on environmental security behind the scenes, or forced it to be relabelled as disaster relief': Geoffrey Dabelko, 'An Uncommon Peace: Environment, Development, and the Global Security Agenda', *Environment* 50 (2008), p. 39. Notably, this view was shared by at least some of the participants of a workshop that discussed a draft version of this book at the Woodrow Wilson Center in late June 2008. An example of continuous work on environmental security issues under the Bush administrations would be USAID's Office of Conflict Management and Mitigation's commissioning of toolkits for conflict prevention, some of which addressed the linkage of environment and violent conflict.

[134] Author telephone interview with Carole Dumaine, Deputy Director for Energy and Environmental Security in the Office of Intelligence and Counterintelligence, US Department of Energy, 20 November 2008.

position to say something about the value of any given desecuriti-sation. Thus the idea that desecuritisation can lead to both politicisa-tion as well as depoliticisation throws up questions about the moral worth of desecuritisation more generally. As a result, we can no longer maintain – as the Copenhagen School does – that desecuritisation is a necessarily positive process. The case of US environmental security shows that although what was done in the name of that policy under the Clinton administrations was a far cry from the securitising move, it was rather a lot better than what followed under the Bush adminis-trations' desecuritisation. The point is that we as security analysts have a responsibility when we speak up for desecuritisation. Crudely put, we cannot let the Bush administrations simply get away with it. With that in mind the next chapter proposes a way in which, in the environmental sector of security, different types of desecuritisation can be morally evaluated.

6 | The moral evaluation of environmental security

Introduction

The Copenhagen School holds that securitisation and desecuritisation should be evaluated in terms of their outcomes. At the same time they think that the two processes will always lead to the same two outcomes, namely, de-democratisation or depoliticisation in the case of securitisation and politicisation in the case of desecuritisation, leaving them to suggest that securitisation is a morally wrong and desecuritisation a morally right process. The case studies of the US environmental security policies under the Clinton and Bush administrations, however, show (1) that not all securitisations are the same, but that securitisations differ in terms of who or what they benefit, ergo in their outcomes; and (2) that desecuritisation does not always lead to politicisation, but that it can lead to depoliticisation instead. If we take the Copenhagen School's assertion that securitisations and desecuritisations should be evaluated in terms of their outcomes seriously, then it matters that a certain process does not always have the same outcome. In other words, the Copenhagen School's evaluation of either process is incomplete, not to say flawed. In this chapter I seek to provide a moral evaluation of environmental security (this includes desecuritisation) that accounts for the varied outcomes of either process; indeed, I make the *consequences* of either process the hallmark of moral evaluation. In moral philosophy, any approach of this sort is known as a consequentialist approach to ethics.

Consequentialism

Consequentialist theories hold that 'the way to tell whether a particular choice is the right choice for an agent to have made is to look at the relevant consequences of the decision: to look at the relevant effects

174

of the decision on the world'.[1] This is a powerful principle. It is such that even non-consequentialists cannot abstain from considering consequences in their recommendations of what is the morally right action. In the words of John Rawls, 'all ethical doctrines worth our attention take consequences into account in judging rightness. One which did not would be simply irrational, crazy'.[2] While consequences thus feature in all types of moral reasoning, consequentialism's second dictum, whereby the right action is the one that maximises the best consequences, does not. Rawls, for instance, objects to consequentialism because it lacks an independent definition of the good. For Rawls, a consequentialist society which is 'rightly ordered, and therefore just, when its major institutions are arranged so as to achieve the greatest net balance of satisfaction summed over all the individuals belonging to it'[3] cannot possibly be just. This is because such a society does not take into account 'how this sum of satisfactions is distributed among individuals',[4] potentially allowing for the ever-increasing gain of the few, whilst tolerating the misery of the many. Another prominent objection against consequentialism is that it may require individuals to do horrible things (such as murder or torture) if such action maximises the greater aggregate good (for example the prevention of further murders).[5] This objection rests on the conviction that certain goods are indispensable; for example, an innocent person's right not to be killed, even if this would prevent the dying of yet greater numbers of innocent persons.[6] These are just two of the more prominent objections against consequentialism. Consequentialists have engaged with these objections in considerable detail, and the debate whether the right account of morality is consequentialist is one of the most elementary debates in moral philosophy. One of the most compelling arguments in favour of consequentialism is that value promotion is the most logical

---

[1] Philip Pettit, 'Introduction', in Philip Pettit (ed.), *Consequentialism* (Aldershot: Dartmouth Press, 1993), p. xiii.

[2] John Rawls, *A Theory of Justice* (Cambridge MA: Harvard University Press, 1971), p. 30.

[3] *Ibid.* p. 22.

[4] *Ibid.* p. 26.

[5] Thomas Nagel, 'War and Massacre', in Samuel Scheffler (ed.), *Consequentialism and its Critics* (Oxford: Oxford University Press, 1988), pp. 52ff.; Elizabeth Anscombe, 'Modern Moral Philosophy', *Philosophy* 33 (1958), p. 10.

[6] James Griffin, 'The Human Good and the Ambitions of Consequentialism', *Social Philosophy and Policy* 9 (1992), p. 120.

way of acting if we wish to improve the world. In the words of the consequentialist philosopher Tim Mulgan:

The simplest way to motivate Consequentialism is to see it as developing the thought that morality or moral action should be concerned with making the world a better place. At the extreme, this argument takes Consequentialism to be true by definition. [. . .] A more modest formulation is that, while it is possible to imagine Non-Consequentialist moral theories, the most rational way to respond to any value is to promote it.[7]

Consequentialists define morality in terms of good consequences. Consequentialism is a theory of the right; it does not tell us what *good consequences* are. As the contemporary American philosopher Peter Railton puts it: 'one has adopted no morality in particular even in adopting consequentialism unless one says what the good is'.[8] Consequentialists hold that an act is morally right when it can be judged according to an agent-neutral value.[9] Agent-neutrality is the property of those values that 'can be articulated without reference back to the valuer'.[10] In more detail still:

If I value a prospect for the increase of happiness it promises, or even for a particular effect it will have, say on planet Earth, then I value it agent-neutrally. If I value it for the benefits it will have for me or mine, or for the fact that it will keep my hands clean, or for any reason of that self-referential kind, then I value it agent-relatively. The theory of the good, or the theory of the valuable, refers to the theory of what ought to be agent-neutrally valued.[11]

One of the most fundamental units of value promoted by consequentialists is the well-being of humanity in general. Theories that regard

---

[7] Tim Mulgan, *The Demands of Consequentialism* (Oxford: Clarendon Press, 2001), pp. 13–14.

[8] Peter Railton, 'Alienation, Consequentialism and Morality', reprinted in Samuel Scheffler (ed.), *Consequentialism and its Critics* (Oxford: Oxford University Press, 1988), p. 108.

[9] Some moral philosophers suggest that the distinction between consequentialist and non-consequentialist theories can be drawn in terms of whether a theory is agent-neutral or agent-relative. See, in particular, David McNaughton and Piers Rawling, 'Honoring and Promoting Values', *Ethics* 102 (1992), pp. 835–43. For a critique see Douglas W. Portmore, 'Can an Act-consequentialist Theory be Agent Relative?' *American Philosophical Quarterly* 38 (2001), pp. 363–77.

[10] Philip Pettit, 'Introduction', pp. xiii–xv.

[11] *Ibid.* p. xv.

well-being as the highest value are so-called welfarist theories,[12] although proponents predictably disagree about its precise meaning. Away from moral philosophy well-being is often understood as the condition achieved when basic needs necessary for human survival (food, shelter, water etc.) are being met. The problem with the basic needs account of well-being, however, is that basic need fulfilment is not enough to make for well-being, which is evident from the fact that some people willingly choose to satisfy their *desires* at the expense of basic needs, provided that certain basic needs are met.[13] Given this, some moral philosophers view well-being not as the satisfaction of basic needs, but instead as the satisfaction of desires. Such accounts are, however, subject to a number of problems. Actual-desire accounts of well-being, where well-being is achieved simply when people get what they want, break with the obligation incentive of needs. It is simply not clear why anyone – this includes states – should give people whatever takes their fancy, especially considering that fancies might change.[14] More sophisticated desire-satisfaction theories are therefore based not on actual desires determined by people's expectations, but rather on 'desires that I would have only if I were properly informed, reflective, and rational, or they require that my desires not result from conditioning or from some undesirable psychological mechanism'.[15] Well-being then 'is the fulfilment of informed desire'.[16] Considering, however, that people disagree over the content of informed desires there are problems with this approach too. The same criticism can also be applied to so-called objective-list theories that regard human well-being as consisting of certain objectively valuable (and worthless) things, while holding that 'these things are good (or bad) for a person regardless of whether the person desires or enjoys them'.[17] In short, while all accounts of well-being have something going for them, not one of them offers a sufficient explanation of well-being. The most

[12] Roger Crisp, 'Well-Being', in Edward N. Zalta (ed.), *The Stanford Encyclopedia of Philosophy* (2008 revised edition), at http://plato.stanford.edu/ entries/well-being/ [1/2009].
[13] James Griffin, *Well-Being: Its Meaning, Measurement and Moral Importance* (Oxford: Clarendon Press, 1986), pp. 40–7.
[14] *Ibid.* pp. 47–51.
[15] William H. Shaw, *Contemporary Ethics: Taking Account of Utilitarianism* (Oxford: Blackwell Publishers, 1999), p. 55.
[16] Griffin, *Well-Being*, p. 75.
[17] Shaw, *Contemporary Ethics*, p. 56.

plausible account of well-being then appears to lie in a combination of all these theories. This much has been suggested by Oxford philosopher Derek Parfit, in a passage that it is worth citing at some length:

[T]he value of the whole may not be the mere sum of the value of its parts. We might then claim that what is best for people is a composite. It is not just their being in the conscious state that they want to be in. Nor is it just their having knowledge, engaging in rational activity, being aware of true beauty, and the like. What is good for someone is neither just what Hedonists claim, nor just what is claimed by Objective-List Theorists. We might believe that if we had *either* of these, *without the other*, what we had would have little or no value. We might claim, for example, that what is good or bad for someone is to have knowledge, to be engaged in rational activity, to experience mutual love, and to be aware of beauty, while strongly just wanting these things. On this view, each side in this disagreement only saw half the truth. Each put forward as sufficient something that was only necessary. Pleasure with many other kinds of object has no value. And, if they are entirely devoid of pleasure, there is no value in knowledge, rational activity, love or the awareness in beauty. What is of value, or is good for someone, is to have both; to be engaged in these activities, and to be strongly wanting to be so engaged.[18]

It is often argued that a functioning natural non-human environment (henceforth 'environment') is essential to human well-being. Anthropocentric environmental ethicists, for example, hold that the environment is of instrumental value, meaning that the value of the environment resides in its utility to human ends only. At the most basic level this means that human beings cannot exist – at least in the way they currently do and have always done – without air to breathe, water to drink and an accommodating climate. At a more refined level it means that human well-being is closely connected to, for instance, an appreciation for natural beauty or the pleasure derived from a country walk.[19] The view that the environment is of instrumental value to human ends is the most commonly accepted view of environmental value. If we accept this view, then, if our concern is human well-being, two things follow: (1) regardless of whether or not individuals are aware of it, care for the environment is already in their own interest, and (2) care for the environment needs to take a specific form, in so far

[18] Derek Parfit, *Reasons and Persons* (Oxford: Clarendon Press, 1984), pp. 501–2.
[19] Compare with Peter Singer, *Practical Ethics*, 2nd edition (Cambridge University Press, 1993), p. 272.

as it ought to 'be very carefully managed for *human benefit*'.[20] The concept of 'benefiting' has already played a major role in this book; in the following it will be vital for evaluating environmental security.

The moral evaluation of environmental security

My point of departure is the idea that if, as under securitisation theory, security is about survival then, if the aim is moral evaluation, the logical follow-on question must be, *who or what should survive?* Proponents of environmental security tend to focus upon the following three candidates for survival: (1) the nation state (this includes its security institutions), (2) human beings (civilisation), and (3) ecosystems (the biosphere). Corresponding to these three, in turn, we can say that there exist three broad categories of environmental security: (1) environmental security as national security, (2) environmental security as human security, and (3) ecological security. Whilst environmental security as national security, which encompasses both defence environmental security and the idea that environmental variables can play a role in violent conflict, has been discussed in the context of the case studies, the other two have not yet been mentioned.

Proponents of environmental security as human security argue that because environmental threats are oblivious to territorial boundaries, true environmental security can only be achieved if the issue is moved out of the traditional state-centric threat and defence nexus. Unlike what many Critical security theorists suggest, this is not necessarily the same as saying that states have become redundant as providers of human security. To the contrary, some proponents of this approach maintain that 'human security *cannot* be separated from the operation of states [and that states] are critical to providing economic opportunities, creating and providing measures to protect people when livelihoods contract'.[21]

[20] Clare Palmer, 'An Overview of Environmental Ethics', in Andrew Light and Holmes Rolston III (eds.), *Environmental Ethics: An Anthology* (Oxford, Blackwell, 2003), p. 18 (emphasis added).
[21] Jon Barnett and W. Neil Adger, 'Environmental Change, Human Security, and Violent Conflict', in Richard A. Matthew, Jon Barnett, Bryan McDonald and Karin L. O'Brian (eds.), *Global Environmental Change and Human Security* (Cambridge MA: MIT Press, 2009), pp. 128–9 (emphasis added); see also Rita Floyd, 'Human Security and the Copenhagen School's Securitization Approach', *Human Security Journal* 5 (2007), p. 44.

Proponents of environmental security as human security focus on issues such as ecological interdependence, human rights, on the impact of globalisation and on the effect of northern consumption patterns on the Global South.[22] For them the nature of the threat stems from the dangers of long-term environmental degradation, such as global warming, species extinction, pollution of air and water, loss of biodiversity and ozone depletion, that are non-violent in character. This type of environmental security has been usefully described as: 'The process of peacefully reducing human vulnerability to human-induced environmental degradation by addressing the root causes of environmental degradation and human insecurity'.[23]

Environmental security as human security is most prominently advocated by the UN, whose 1994 Human Development Report laid the groundwork for the concept of human security. That report also included environmental security. More recently, the UN has been active on the subject of climate change as a security issue. The 2007–08 Human Development Report, with the title *Fighting Climate Change: Human Solidarity in a Divided World*, focuses on future generations and on the world's poorest people, the latter of which will not only be the first to be adversely affected by climate change, but, largely because they lack necessary adaptation, will also be affected the worst. It warns that climate change is a 'massive threat to human development [which] in some places is already undermining the international community's efforts to reduce extreme poverty'.[24] Notably, this report foresees no role for traditional security institutions such as the military in its suggested solutions; instead it stresses that only global cooperative measures will bring down emissions.

Proponents of ecological security are most commonly members of the various green activist and lobbyist non-governmental organisations but can also be found in Green Parties. Ecological security is based on the ideology of deep green ecology, and focuses on non-violent environmental issues such as biodiversity loss and climate change.

[22] See, for example, Simon Dalby, *Environmental Security* (Minneapolis: Minnesota Press, 2002).

[23] Jon Barnett, *The Meaning of Environmental Security: Ecological Politics and Policy in the New Security Era* (London: Zen Books, 2001), p. 129.

[24] UN, Human Development Report 2007/2008, *Fighting Climate Change: Human Solidarity in a Divided World* (New York: United Nations Development Programme, 2007), pp. v, vi.

There are two main components to deep green ecology: self-realisation, meaning the identification of the 'self' with a larger organic entity; and bio-centric equality, the principle whereby all species are equal.[25] It follows from this that humans are secured under the concept of ecological security only as part of a larger ecosystem, and only in the same way as all other species.[26] Deep green thought argues that global environmental change does not happen irrespective of international power structures, but rather international power structures actively create global environmental change.[27] These power structures can be divided into three categories: social/cultural (values and patriarchy), political (the state-system), and economic (globalisation, hegemony and capitalism). Ecological security then can only be achieved when these power structures are dismantled, or at least changed to embrace ecologically sustainable modes of production. The provider of ecological security then is humanity at large, as only humanity is capable of changing existing patterns of ecological destruction and embracing an approach where ecology is paramount.

From the point of view of well-being consequentialism, *only* environmental security as human security is morally permissible because only in this approach is the individual the guaranteed beneficiary of the securitisation – that which should survive. Ecological security starts from altogether different premises. Here the environment is valued not because it has instrumental value to human ends but rather because ecosystems have intrinsic value 'independent of that of their component individuals'.[28] The crucial point here is that humans have no more worth than the ecosystem as a whole or, in shallower green formulations, than other living or sentient creatures. Since a vast majority of environmental problems, including climate change, are anthropogenic, humans themselves are the source of the environmental (or better ecological) insecurity. If this is so then it is possible to imagine that a securitisation

---

[25] Bill Devall and George Sessions cited in John S. Dryzek, *The Politics of the Earth* (Oxford: Oxford University Press, 1997), p. 156. See also Barnett, *The Meaning of Environmental Security*, ch. 8.

[26] Dennis Pirages, 'Demographic Change and Ecological Security', *Environmental Change and Security Project Report* (Washington DC: The Woodrow Wilson Center 1997), p. 37.

[27] Matthew Paterson, *Understanding Global Environmental Politics: Domination, Accumulation and Resistance* (Basingstoke: Palgrave, 2001), pp. 35ff.

[28] Robin Attfield, *Environmental Ethics* (Cambridge: Polity, 2003), p. 192.

informed by these ideas could be accompanied by misanthropic extraordinary measures. Indeed, one prominent ecocentric suggests that 'the more misanthropy there is in an ethical system, the more ecological it is, and that the human population should be, in total, about twice that of bears'.[29] Whilst such measures would presumably not target already existing humans, they would negatively affect future generations. This is especially so because population growth is often and controversially attributed a pivotal role in the proliferation of environmental problems.[30]

From a position of well-being consequentialism, the problems with environmental security as national security are just as severe as they are with ecological security; they are, however, of an entirely different making. Here human beings are secured only in so far as they are part of a state that prioritises environmental problems. Whilst this is not necessarily problematic – at least not for the people living in that state – the case of the Clinton administrations' environmental security policy shows that this prioritisation does not automatically mean that the environmental problems of our time are addressed in security mode. Instead, environmental security might mean little more than the military cleaning up their contaminated bases or, more crudely put, the military finally complying with existing environmental legislation. An additional problem with environmental security as national security is that proponents locate the source of the threat with the environment itself.[31] Environmental degradation is of interest only in so far as it may threaten military readiness and/or leads to violent conflict. As such, environmental security as national security is squarely part of the old, traditional threat–defence nexus; it 'is simply an additional component to pre-existing notions of security. The referent of security remains the state, while only the causes of insecurity have changed – from military enemies to environmental degradation'.[32]

---

[29] Palmer, 'An Overview of Environmental Ethics', p. 24. See also the chapter by J. Baird Callicott in the same volume. Callicott is the originator of the bear–human population balance statement.

[30] Betsy Hartmann, 'Rethinking the Role of Population in Human Security', in Richard A. Matthew, Jon Barnett, Bryan McDonald and Karin L. O'Brian (eds.), *Global Environmental Change and Human Security* (Cambridge MA: MIT Press, 2009), pp. 192–214.

[31] Dalby, *Environmental Security*, p. 22.

[32] Paterson, *Understanding Environmental Politics*, p. 20.

Whilst there can be no question that the issues that have been addressed by proponents of environmental security as national security are important, addressing these issues as part of traditional national security is not advisable. The reasons for this are different for the greening of defence approach and for the various environment-conflict theses. In the case of defence environmental security the security equation is made simply by virtue of traditional security institutions addressing environmental issues. In this understanding then it is *not* the nature of the issue that marks something out as a security issue; instead something (the environment) becomes a security issue by virtue of being handled by a traditional security institution. As such this approach practically invites agent-benefiting securitisations.

Although, in the case of potential violent environmental conflict the security equation is readily given, securitisation is not the way forward here. Whilst it is possible to address environmental problems as part of the old threat and defence nexus, such problems are very different in nature to the threats usually found there. Above all, at their core, all environmental problems are *collective* problems. Environmental conflict, for example, might at first sight appear to play out in a fairly limited local realm only; often, the causes of such conflict lie with the dynamics of the global political economy.[33] What is more, leaving aside the issue of moral obligation to 'save strangers',[34] whilst violent conflict where environmental stress is a contributing variable is at present very much restricted to the developing world, conflict-induced environmental migration concerns all of us.[35] In short, collective environmental problems can only be addressed through cooperation, and as the classic security dilemma readily shows, national security severs cooperation.[36] In the short run, some wealthy states may be in the position to secure themselves against the side effects of environmental problems. For example, stricter border controls can be

---

[33] Nancy L. Peluso and Michael Watts (eds.), *Violent Environments* (New York: Cornell, 2001).

[34] Nicholas J. Wheeler, *Saving Strangers: Humanitarian Intervention in International Society* (Oxford: Oxford University Press, 2000).

[35] Thomas Homer-Dixon, 'On the Threshold: Environmental Changes as Causes of Acute Conflict', *International Security* 16 (1991), p. 113.

[36] Marc A. Levy, 'Is the Environment a National Security Issue?' *International Security* 20 (1995), pp. 47ff.

introduced in an effort to keep environmental migrants at bay,[37] whilst scarce resources can, if necessary, be obtained by force. The point is rather that none of these actions address the root causes of environmental problems – for example, underdevelopment and poverty; they simply deal with some of the anticipated adverse consequences of environmental insecurity for the nation state, whilst environmental problems multiply.[38] In the long run therefore, environmental security as national security is counterproductive.

In summary, neither the consequences of environmental security as national security nor those of ecological security are conducive to human well-being. The former benefits the state and its security institutions only, while the latter benefits the ecological system as a whole. Though human beings are considered part of this, they are also at the source of the threat and need dealing with – ergo the consequences for human life, especially future generations, may be detrimental.

Only environmental security as human security directly benefits human beings. Unlike environmental security as national security, it does not aim at merely containing some of the anticipated adverse consequences of environmental change and effectively securing one group of people before either (a) the ill-functioning environment and/ or (b) the primary victim of environmental change – environmental migrants. Instead it seeks to address the *root causes* of environmental change through (global) *cooperative measures* with the ultimate aim of establishing a healthy and functioning environment for all of us. Importantly, for human security, the environment is neither regarded as a 'villain' negatively affecting national security nor as equally valuable to human existence. Instead, this approach stresses the fundamental *interdependence* between well-being and a healthy environment. It is as such a morally right process, on account of its being the only policy agenda properly conducive to human well-being. Environmental security as national security and ecological security are, by contrast, viewed as morally wrong processes from here on in.

Let us now turn to the issue of desecuritisation. I have in this book argued that desecuritisation can take two specific forms, desecuritisation

---

[37] For example, India is currently building a 2,500-mile long fence that will keep out Bangladeshi flood victims and environmental migrants.

[38] Tellingly, there now exists evidence which suggests that 'human insecurity increases the risk of violent conflict'. Barnett *et al.*, 'Environmental Change, Human Security, and Violent Conflict', p. 128.

as politicisation and desecuritisation as depoliticisation. Since these different types of desecuritisation exist, it is plausible to suggest that not all desecuritisations are morally equal. If desecuritisation only ever takes two distinct forms, then it is plausible to suggest that in the environmental sector of security one of these must be morally right and the other must be morally wrong. In order to uncover which type of desecuritisation is morally permissible in this sector, I begin by considering the question: should global environmental issues be matters of high politics? If this question is answered in the affirmative then, in the environmental sector of security, desecuritisations as politicisations are morally permissible and desecuritisations as depoliticisations are not. It is important to note that, by posing the above question, I am not asking whether global environmental issues relate, or should relate, to a state's national interest, which is one definition of high politics. Instead, I am interested in whether or not they should be dealt with by '*formal office holders*', i.e. national leadership.[39] High politics therefore corresponds to my reading of politicisation as used throughout this book.

Informed by the idea that the environment is a necessary requirement for human well-being, the question whether or not global environmental issues should be dealt with by those in power can easily be answered with a 'yes'. Whilst it is possible that instances of depoliticisation – market-led solutions, for example – can sometimes provide solutions to specific environmental problems, in the vast majority of cases political solutions to environmental problems are necessary.[40] In almost every instance, action on environmental problems can only be ensured as part of a political solution and, most notably, global environmental regimes.

Consider as an example of this the issue of global climate change. Since the publication of the IPCC's fourth assessment report on climate change in early 2007 we have something very close to scientific consensus on (a) that global climate change is occurring, and also on (b) that it is anthropogenic in nature. Since awakening to this news, countless solutions to curbing the growth of further emissions have been put forward. Some of these lie with technology, such as carbon

[39] See Graham Evans and Jeffrey Newnham, *The Penguin Dictionary of International Relations* (London: Penguin Books, 1998), p. 225, for more information on high politics and elite structures.
[40] Anthony Giddens, *The Politics of Climate Change* (Cambridge: Polity, 2009), pp. 5, 91.

capture and storage or with the development of hybrid vehicles and alternative (green) fuels. Other solutions lie with economics through various forms of emissions trading, for instance, cap and trade programmes and/or joint implementation. There has also been a combination of technological and economic solutions in that – in some areas at least – renewable energies are now seen as important business opportunities. In the eastern part of Germany, for example, the solar industry is reviving derelict industry in the outskirts of Leipzig and Frankfurt (Oder). Although all of these solutions are worthwhile, by themselves these solutions will be more attractive to those already committed to climate action. What is needed is a *political* solution that enforces climate action by everyone. What is needed is a post-Kyoto global environmental treaty that sets mandatory targets so that emissions will rise to, at the very most, 450ppm CO2eq (from currently 430ppm CO2eq). This would translate to a 2 to 3.5 °C increase in the average global temperature.[41] Since global environmental treaties can be signed into action by power-holders alone, there can be little doubt about the fact that global environmental issues should be matters of politics. In the environmental sector of security, we can therefore conclude that, in almost all cases, desecuritisation as politicisation is morally right, whereas a desecuritisation as depoliticisation is morally wrong.

Conclusion

The categories developed in this chapter of morally right and morally wrong securitisation and desecuritisation in the environmental sector of security are useful because they enable the securitisation analyst to offer concrete advice to potential securitising actors on how to securitise or not to securitise. On the basis of this moral evaluation I am now in the position to denounce the Clinton administrations' environmental security policy as a morally wrong securitisation because, although the administrations employed some of the rhetoric of environmental security as human security, human beings were *not* the beneficiary of environmental security. Instead, as Chapter 4 has clearly shown, the beneficiary of this policy was the securitising actor.

---

[41] Gabrielle Walker and Sir David King, *The Hot Topic: How to Tackle Climate Change and Still Keep the Lights On* (London: Bloomsbury, 2008), p. 96.

My evaluative framework enables me to qualify further my reservations about the Copenhagen School's one-sided view of desecuritisation. I am now not only able to say that desecuritisation can lead to politicisation as well as depoliticisation, but also that, in the environmental sector of security, the latter is morally wrong. It is on this basis that I am now able to condemn as morally wrong what, in particular, the Bush administrations have been doing, or rather not doing, regarding environmental security.

7 | Conclusion

It was the purpose of this book to revise the Copenhagen School's influential securitisation theory so that it both allows insights into the intentions of securitising actors, and permits the moral evaluation of securitisation and desecuritisation in the environmental sector of security. Without doubt, being able to account for why an actor securitised is one of the most important and interesting issues in security analysis. Every time something momentous happens in world politics (for instance, the events of 9/11, or the 2003 invasion of Iraq) friends and family – suddenly remembering what it is that we do for a living – will ask us, why did this happen? Why did they (the terrorists, the states, the statesmen etc.) do this? Surely, they would be astounded if we told them both that security analysts cannot tell them anything meaningful regarding intentions and that we content ourselves with simply analysing who did what exactly, under what circumstances and to what effect. Without doubt they would think that we are only offering half of an analysis, and that one of the most interesting but also challenging tasks to face us is one from which we shy away.

The Copenhagen School's belief that we cannot know anything about a securitising actor's intentions rests, on the one hand, on the conflation of the philosophically distinct concepts of 'motives' and 'intentions', and, on the other, on an aversion to causal theorising, as associated with positivist analysis. It is for this reason that securitisation theory is often described as a constitutive theory, concerned with, in Alexander Wendt's terminology, not causal 'why' questions, but rather constitutive 'how possible' questions.[1] As Milja Kurki has recently shown, however, this distinction, together with the post-positivist aversion to causal analysis, rests on a narrow empiricist understanding of cause, whereby the term refers only to observable regulatory-deterministic

---

[1] Alexander Wendt, *Social Theory of International Relations* (Cambridge University Press, 1999), pp. 78ff.

188

causes.[2] In addition, she convincingly shows that, regardless of their theoretical position, all IR theorists engage in causal analysis. For instance, even 'when poststructuralists highlight the role of discourse or theories in "constituting" social life, they do so because these discourses or theories, through constituting agents' perceptions and reasoning, have "consequences" for how agents perceive the world, themselves, others and, hence, their actions'.[3] To close the artificial divide between causal and constitutive IR theorising, Kurki advances an alternative notion of cause borrowed from scientific realism, a definition that refers to 'all those things that bring about, produce, direct or contribute to states of affairs or changes in the world',[4] including discourses, intentions and speech acts. My redefinition of securitisation theory works very much in the spirit of Kurki's ambitious proposal. Her work convincingly shows, I think, that causal analysis, including the analysis of intentions as causes, is perfectly compatible with securitisation theory. As regards consequences, they are, as I have tried to show, already implicitly part of securitisation analysis (as evinced, to reiterate, by the fact that Wæver's claim that the securitisation analyst has a responsibility to point to desecuritisation is only comprehensible in the light of his anticipated consequences of desecuritisation, namely politicisation). I would suggest that reconciling securitisation theory with intentions and consequences in this way not only makes securitisation theory more coherent, it also makes it much more attractive both to those security analysts positioned at the mainstream end of security studies (whose sole concern is already causal analysis), and to those who (regardless of their disciplinary position) would wish to look to security studies not just for insights as regards the practice of security but also for some clue as to how we might guide such practices in the future.

My comprehensive analysis of US environmental security, including the moral evaluation of environmental security, is particularly timely considering that environmental security is destined to return to US national security, if under the label 'climate security'. Thus, President

[2] Milja Kurki, *Causation in International Relations: Reclaiming Causal Analysis* (Cambridge University Press, 2008), p. 6.
[3] Milja Kurki, 'Causes of a Divided Discipline: Rethinking the Concept of Cause in International Relations', *Review of International Studies* 32 (2006), p. 212.
[4] *Ibid.* p. 202.

Obama regards climate change as a security issue.[5] The US intelligence community believes that 'global climate change will have wide-ranging implications for US national security interests over the next 20 years'.[6] Secretary of State Hillary Clinton declares that climate change is 'more than a scientific phenomenon. It's a political challenge, it's an economic force, it's a *security threat*, and a moral imperative'.[7] And the US Department of Defense considers climate and environmental pressures uncertainties impacting on the *future* strategic environment.[8] Although there is no consensus on the meaning of climate security,[9] in the US state-centric approaches to climate security are dominant and some of the most vocal proponents of 'climate security' have close ties with the military. In fact, some of the most prolific voices in the climate security debate are, with Sherri Goodman, Anthony Zinni, James Woolsey and Leon Fuerth, some of the same people that were at the centre of environmental security under the Clinton administration. Together with ten retired Army and Navy Generals, Goodman and Zinni produced a 2007 report with the title *National Security and the Threat of Climate Change* for the Center for Naval Analyses (the CNA Corporation), a non-profit research organisation, of which Goodman is the acting General Counsel. The report sits squarely with the old defence environmental security approach. On the one hand, it highlights the role of the defence and intelligence community in this new area, for instance, through environmental disaster prevention and relief. It also acknowledges the possibility of violent conflict due to resource shortages. On the other hand, the report focuses on the consequences of climate change for the US national security establishment; for example, there is concern for DOD installations situated in areas

[5] Steve Holland, 'Obama Says Climate Change a Matter of National Security', *Reuters*, 9 December 2008, at www.reuters.com/article/environmentNews/idUSTRE4B86R920081209 [1/2009].

[6] Thomas Fingar, '2008 National Intelligence Assessment on the National Security Implications of Global Climate Change' (Washington DC: US Senate, House Permanent Select Committee on Intelligence and House Select Committee on Energy Independence and Global Warming, 2008).

[7] Hillary Rodham Clinton, 'Remarks at the State Department's "Greening Diplomacy" Earth Day event', 22 April 2009, at www.state.gov/secretary/rm/2009a/04/122064.htm [5/2009] (emphasis added).

[8] US Department of Defense, '2008 National Defense Strategy' (Washington DC: Department of Defense, June 2008), pp. 4, 5.

[9] Rita Floyd, 'The Environmental Security Debate and its Significance for Climate Change', *The International Spectator* 43 (2008), pp. 51–65.

that would be affected by sea level rise. The message is clear: climate security should be recognised as a threat to national security and the military has an important role to play in it. Fuerth and Woolsey are with *The Age of Consequences: The Foreign Policy and National Security Implications of Global Climate Change*, a 2007 report released by the Center for Strategic and International Studies and the Center for a New American Century, part of a similar multi-authored effort. This report identifies a strong linkage between climate change and terrorism and posits the transformation of America's energy economy as the solution to both problems.[10]

Although these reports are not part of official government policy their importance should not be underestimated; at least some of what they propose could very well become part of the Obama administration's security agenda. The report sponsored by the CNA Corporation, for instance, is taken very seriously by Democrats in Washington. Goodman has testified on the report to the Senate's Committee on Energy and Commerce and the Subcommittee on Energy and Air Quality. The report has been welcomed by Senator John Kerry, who led the Senate's delegation to the UN climate meeting in the Polish city of Poznan in December 2008. Perhaps most significantly of all, the findings of the CNA report played a part in Congress mandating (as part of the National Defense Authorization Act for Fiscal Year 2008) that the National Defense Strategy and the National Security Strategy issued after this act must include the following considerations concerning climate change: '[G]uidance for military planners (A) to assess the risks of projected climate change to current and future missions of the armed forces; (B) to update defense plans based on these assessments, including working with allies and partners to incorporate climate mitigation strategies, capacity building, and relevant research and development; and (C) to develop the capabilities needed to reduce future impacts'.[11]

These developments suggest that there exists a real possibility that 'climate security', like environmental security under the Clinton

[10] Kurt M. Campbell *et al.*, *The Age of Consequences: The Foreign Policy and National Security Implications of Global Climate Change* (Washington DC: Center for Strategic and International Studies and the Center for a New American Century, 2007).
[11] United States House of Representatives, 'National Defense Authorization Act for Fiscal Year 2008' (Washington DC: US Congress), section 931, p. 276.

administrations, will be first and foremost about military readiness. As such, climate security could give those policy-makers with little interest in environmental issues a shield to hide behind, as those reluctant to sign up to fixed carbon emission targets would be able to say, 'We care about the climate so much, we even consider it a matter of national security', whilst doing little above and beyond securing military installations before the ill-effects of climate change. History, however, must not necessarily repeat itself and I could easily be accused of painting an overly bleak picture here. The point of doing this is that security analysts must realise that 'climate security' may mean a lot less than one might initially think and that it might not necessarily be a good thing.

Further research

The moral evaluation of security in one sector of security potentially has significant implications for security studies more generally. It suggests that the focus of security studies could shift from an almost exclusive focus on 'how do actors securitise', to include a focus on *when and how to securitise*. Such a move would correspond with the message of a recent mainstream constructivist edited book that suggests that international relations theorists should focus on 'how should we act'.[12] This book further suggests that constructivist analysis lends itself well to normative theorising, because constructivists hold that a responsible answer to the question 'how should we act?' 'depends not just on what one judges as right in the abstract, but also on what one may have some reasonable expectation of working, and thus prescribing as a course of action or judgement'.[13] In other words, normativity must be limited to what is possible. In security relations we can determine what is possible only if we free ourselves from what we would ideally want to be the case, and focus on how security *works* instead. This is where securitisation theory comes in; it provides us with useful insights into what is possible. Only after we have answers to this

---

[12] Christian Reus-Smit, 'Constructivism and the Structure of Moral Reasoning', in Richard M. Price (ed.), *Moral Limit and Possibility in World Politics* (Cambridge University Press, 2008), pp. 64ff.

[13] Richard M. Price, 'Moral Limit and Possibility in World Politics', in Richard M. Price (ed.), *Moral Limit and Possibility in World Politics* (Cambridge University Press, 2008), pp. 6–7.

question can we, in a second step, move on to normative territory and, in the light of *what is possible*, suggest feasible scenarios for *what ought to be* the case.

Although my study of the environmental sector of security does not hold answers to the question 'how should we securitise' for the remaining sectors of security (indentified by the Copenhagen School as military, political, societal and economic sectors of security), it does hold a number of valuable insights for anyone interested in pursuing such a study. These are: (1) not all securitisations are the same, they differ in terms of *who or what* they benefit; (2) desecuritisation can lead to politicisation but also to depoliticisation; (3) consequentialism is a constructive theory of right; (4) human well-being is a useful unit of value. The fourth of these points is the most challenging bit. Whilst in the environmental sector of security it is comparatively easy to argue that the environment is a necessary requirement of well-being, what else in the other sectors could be said to have that same status? Given that different scholars inevitably provide different answers to this question, there is huge potential for the development of different competing yet also complementary research projects of a sort that would move security analysis from its present concern with 'how do actors securitise?', towards a consideration of the question 'how should actors securitise?'

Bibliography

Aadland, Anders, 'Leaked Memo to Garrison Staff', Public Employees for Environmental Responsibility, Washington DC, 11 May 2004

Alker, Hayward, 'Emancipation in the Critical Security Studies Project', in Ken Booth (ed.), *Critical Security Studies and World Politics*, Boulder, Lynne Rienner, 2005, 189–214

Allenby, Braden, 'New Priorities in US Foreign Policy: Defining and Implementing Environmental Security', in Paul Harris (ed.), *The Environment, International Relations and US Foreign Policy*, Washington DC, Georgetown University Press, 2001, 45–67

Anscombe, Elizabeth, *Intention*, Oxford, Basil Blackwell, 1957
  'Modern Moral Philosophy', *Philosophy* 33, 1958, 1–19

Aradau, Claudia, 'Security and the Democratic Scene: Desecuritization and Emancipation', *Journal of International Relations and Development* 7, 2004, 388–413

Arendt, Hannah, *The Human Condition*, second edition, Chicago, University of Chicago Press, 1998

Asmus, Ronald, 'The European Security Agenda', in Roland Dannreuther and John Peterson (eds.), *Security Strategy and Transatlantic Relations*, Abingdon, Routledge, 2006, 17–29

Atlas, Terry, 'Tim Wirth Takes on the World of Problems' *Chicago Tribune*, 7 September 1994

Attfield, Robin, *Environmental Ethics*, Cambridge, Polity, 2003

Atwood, Brian J., 'Remarks to the Conference on "New Directions in US Foreign Policy"' at the University of Maryland, College Park', reproduced in *Environmental Change and Security Project Report*, Washington DC, The Woodrow Wilson Center, 1996, 85–8

Austin, John L., *How to Do Things with Words*, New York, Oxford University Press, 1962
  'Speech Acts and Convention: Performative and Constative', in Susana Nuccetelli and Gary Seay (eds.), *Philosophy of Language: The Central Topics*, Lanham MD, Rowman & Littlefield Publishers, 2008, 329–36

Baechler, Günther, 'Why Environmental Transformation Causes Violence: A Synthesis', *Environmental Change and Security Project Report*, Washington DC, The Woodrow Wilson Center, 1998, 24–44

Violence through Environmental Discrimination: Causes, Rwanda Arena, and Conflict Model, Dordrecht, Kluwer Academic Publishers, 1999

Bagge Laustsen, Carsten and Wæver, Ole, 'In Defence of Religion: Sacred Referent Objects of Securitization', *Millennium: Journal of International Studies* 29, 2000, 705–39

Baldwin, David, 'Security Studies and the End of the Cold War', *World Politics* 48, 1995, 117–41

Balzacq, Thierry, 'The Three Faces of Securitization: Political Agency, Audience and Context', *European Journal of International Relations* 11, 2005, 171–201

Barna, Theodore, 'Written Testimony on Oil Shale and Oil Sands Resources Hearing', Washington DC, Senate Committee on Energy and Natural Resources, 12 April 2005

Barnett, Jon, *The Meaning of Environmental Security: Ecological Politics and Policy in the New Security Era*, London, Zed Books, 2001

Barnett, Jon and Adger, W. Neil, 'Environmental Change, Human Security, and Violent Conflict', in Richard A. Matthew, Jon Barnett, Bryan McDonald and Karin L. O'Brian (eds.), *Global Environmental Change and Human Security*, Cambridge MA, MIT Press, 2009, 119–36

Barringer, Felicity, 'Officials Reach California Deal to Cut Emissions', *New York Times*, 31 August 2006

Bartis, James T., LaTourette, Tom, Dixon, Lloyd, Peterson, D. J. and Cecchine, Gary, *Oil Shale Development in the United States: Prospects and Policy Issues*, RAND Infrastructure, Safety, and Environment Report Prepared for National Energy Technology Laboratory of the US Department of Energy, 2005

Bomberg, Elisabeth and Super, Betsy, 'The 2008 US Presidential Election: Obama and the Environment', *Environmental Politics* 18, 2009, 424–30

Booth, Ken, 'Security and Emancipation', *Review of International Studies* 17, 1991, 313–26

'Security in Anarchy: Utopian Realism in Theory and Practice', *International Affairs* 67, 1991, 527–45

Theory of World Security, Cambridge University Press, 2007

Booth, Ken (ed.), *Critical Security Studies and World Politics*, Boulder, Lynne Rienner, 2005

Borger, Julian, Adam, David and Goldenberger, Suzanne, 'Bush Kills off Hopes for G8 Climate Change Plan', *The Guardian*, 1 June 2007

Bourdieu, Pierre, *Language and Symbolic Power*, Cambridge, Polity Press, 1992

Brundtland, Gro Harlem (ed.), *Our Common Future: The World Commission on Environment and Development*, Oxford, Oxford University Press, 1987

Bureau of Oceans and International Environmental and Scientific Affairs, 'Fact Sheet Millennium Challenge Account', Washington DC, Department of State, August 2002

Bush, George, 'State of the Union 2002', Washington DC, Office of the Press Secretary, 29 January 2002

'Global Development, President Bush Remarks to Inter-American Development Bank', Washington DC, Office of the Secretary of State, 14 March 2002

'Climate Change Fact Sheet', Washington DC, Office of the Press Secretary, 18 May 2005

'President Bush and the Asia-Pacific Partnership on Clean Development', Washington DC, Office of the Press Secretary, 27 July 2005

'Fact Sheet: President Bush Addresses the Nation on Recovery from Katrina', Washington DC, Office of the Press Secretary, 15 September 2005

'State of the Union 2006', Washington DC, Office of the Press Secretary, 31 January 2006

'State of the Union 2007', Washington DC, Office of the Press Secretary, 23 January 2007

'President Bush Participates in Major Economies Meeting on Energy Security and Climate Change; George W. Bush, President Remarks', Washington DC, Department of State, 28 September 2007

'President Bush Discusses Climate Change', Washington DC, Office of the Press Secretary, 16 April 2008

'Fact Sheet: Taking Additional Action to Confront Climate Change', Washington DC, Office of the Press Secretary, 2008, 16 April 2008

'President Bush Discusses Energy', Washington DC, Office of the Press Secretary, 18 June 2008

Buzan, Barry, *People, States and Fear: The National Security Problem in International Relations*, London, Harvester Wheatsheaf, 1983

People, States and Fear: An Agenda for International Security Studies in the Post-Cold War Era, second edition, London, Harvester Wheatsheaf, 1991

Buzan, Barry and Wæver, Ole, 'Slippery? Contradictory? Sociologically Untenable? The Copenhagen School Replies', *Review of International Studies* 23, 1997, 241–50

Regions and Powers: The Structure of International Security, Cambridge University Press, 2003

Buzan, Barry, Wæver, Ole and De Wilde, Jaap, *Security: A New Framework for Analysis*, Boulder, Lynne Rienner, 1998

Campbell, David, *Writing Security*, second edition, Minneapolis, University of Minnesota Press, 1998

Campbell, Kurt M., Gulledge, Jay, McNeill, J. R., Podesta, John, Ogden, Peter, Fuerth, Leon, Woolsey, R. James, Lennon, Alexander T. J., Smith, Julianne, Weitz, Richard and Mix, Derek, *The Age of Consequences: The Foreign Policy and National Security Implications of Global Climate Change*, Washington DC, Center for Strategic and International Studies and the Center for a New American Century, 2007

Carter, Ashton, 'DoD News Briefing on the New Information Sharing Initiative with the Russian Government, a Result of the recent Gore–Chernomyrdin Commission Meeting', Washington DC, Office of the Assistant Secretary of Defense, Public Affairs, 1 February 1996

Center for Defense Information Washington DC, 'The Military and the Environment', *The Defense Monitor* 23(9), 1994, 1–7

Christopher, Warren, 'Secretary Warren Christopher: Address to the World Business Council in San Salvador, El Salvador', Washington DC, US Department of State Dispatch, 27 February 1996

Clifford, Frank, 'Christopher Calls for Emphasis on Resources', *Los Angeles Times*, 10 April 1996

Clinton, Bill, 'Inaugural Address', Washington DC, Office of the Press Secretary, 20 January 1993

'State of the Union 1993', Washington DC, Office of the Press Secretary, 17 February 1993

'President Clinton's Remarks on Earth Day 1993', *Environmental Change and Security Project Report*, Washington DC, The Woodrow Wilson Center, 1995, 50–1

'President Clinton's Remarks to the National Academy of Sciences', *Environmental Change and Security Project Report*, Washington DC, The Woodrow Wilson Center, 1995, 51–2

Clinton, Hillary Rodham, 'Remarks at the State Department's "Greening Diplomacy" Earth Day Event', Washington DC, Department of State, 22 April 2009

CNA Corporation, *National Security and the Threat of Climate Change*, Alexandria VA, CNAC, 2007

Collier, Paul and Hoeffler, Anke, 'Greed and Grievance in Civil War', Working Paper, Oxford, Centre for the Study of African Economies, 2002

Conca, Ken and Dabelko, Geoffrey D. (eds.), *Environmental Peacemaking*, Washington DC, Woodrow Wilson Center Press, 2002

Congressional Record Senate Hearing Legislative Day, Monday 11 June 1990, Washington DC, Senate Records

Connaughton, Jim, Hays, Sharon and Watson, Harlan, 'Press Briefing via Conference Call by Senior Administration Officials on IPCC Report', Washington DC, Office of the Press Secretary, 16 November 2007

Crisp, Roger, 'Well-Being', in Edward N. Zalta (ed.), *The Stanford Encyclopedia of Philosophy*, 2008 revised edition, http://plato.stanford.edu/entries/well-being/

Croft, Stuart, *Culture, Crisis and America's War on Terror*, Cambridge University Press, 2006

Dabelko, Geoffrey D., 'Tactical Victories and Strategic Losses: The Evolution of Environmental Security', unpublished doctoral thesis, Faculty of the Graduate School of the University of Maryland, 2003

'An Uncommon Peace: Environment, Development, and the Global Security Agenda', *Environment* 50(3), 2008, 32–45

Dabelko, Geoffrey D. and Simmons, P. J., 'Environment and Security: Core Ideas and US Government Initiatives', *SAIS Review* 17, 1997, 127–46

Dalby, Simon, *Environmental Security*, Minneapolis, University of Minnesota Press, 2002

Dannreuther, Roland and Peterson, John, 'Introduction: Security Strategy as Doctrine', in Roland Dannreuther and John Peterson (eds.), *Security Strategy and Transatlantic Relations*, Abingdon, Routledge, 2006, 1–16

Defense Closure and Realignment Commission, 'Report to the President', Washington DC, Department of Defense, 1995

Defense Closure and Realignment Commission, 'Final Report to the President', Washington DC, Department of Defense, 2005

Deibert, Ronald J., 'From Deep Black to Green? Military Monitoring of the Environment', *Environmental Change and Security Project Report*, Washington DC, The Woodrow Wilson Center, 1996, 28–32

Derrida, Jacques, *Margins of Philosophy*, Chicago, University of Chicago Press, 1982

Of Grammatology, Baltimore, Johns Hopkins University Press, 1998

De Soysa, Indra, 'The Resource Curse: Are Civil Wars Driven by Rapacity or Paucity?' in Mats Berdal and David Malone (eds.), *Greed and Grievance: Economic Agendas in Civil War*, Boulder, Lynne Rienner, 2000, 113–35

Deudney, Daniel, 'The Case against Linking Environmental Degradation and National Security', *Millennium* 19, 1990, 461–76

Devine, Robert S., *Bush versus the Environment*, New York, Anchor Books, 2004

De Vries, Lloyd, 'Bush Disses Global Warming Report', *CBS News*, 4 June 2002

Diez, Thomas and Higashino, Atsuko, '(De)Securitisation, Politicisation and European Union Enlargement', unpublished paper presented at the 29th British International Studies Annual Conference, University of Warwick, 2004

Dobriansky, Paula and Connaughton, James, 'Briefing: US Participation in the Asia-Pacific Partnership on Clean Development and Climate Change', Washington DC, Department of State, 6 January 2006

Dobriansky, Paula, Connaughton, James and Watson, Harlan, 'December 13 Press Conference by the US Delegation', Bali, Indonesia, Department of State, 13 December 2007

Dryzek, John S., *The Politics of the Earth*, Oxford, Oxford University Press, 1997

DuBois, Raymond, 'Pentagon is a Good Steward of the Environment', *USA Today*, 27 October 2004

Durant, Robert F., *The Greening of the US Military: Environmental Policy, National Security, and Organizational Change*, Washington DC, Georgetown University Press, 2007

Dycus, Stephen, *National Defense and the Environment*, Hanover, University Press of New England, 1996

Eriksson, Johan, 'Observers or Advocates? On the Political Role of Security Analysis', *Cooperation and Conflict* 34(2), 1999, 311–30

European Academies Science Advisory Council, *A Study on the EU Oil Shale Industry: Viewed in the Light of the Estonian Experience*, report by EASAC to the Committee on Industry, Research and Energy of the European Parliament, 2007

Evans, Graham and Newnham, Jeffrey, *The Penguin Dictionary of International Relations*, London, Penguin Books, 1998

Evans, Mary Margaret, Mentz, John W., Chandler, Robert and Eubanks, Stephanie L., 'The Changing Definition of National Security', in Miriam R. Lowi and Brian R. Shaw (eds.), *Environment and Security: Discourses and Practices*, New York, St Martin's Press, 2002, 11–31

Fairchild, C. J., 'Does Our Nation's Security Strategy Address the Real Threats?' (1989), www.globalsecurity.org/military/library/report/1989/FCJ.htm

Fingar, Thomas, '2008 National Intelligence Assessment on the National Security Implications of Global Climate Change', Washington DC, US Senate, House Permanent Select Committee on Intelligence and House Select Committee on Energy Independence and Global Warming, 25 June 2008

Firestone, David, 'Drilling in Alaska: A Priority for Bush Fails in the Senate', *New York Times*, 20 March 2003

Floyd, Rita, 'Human Security and the Copenhagen School's Securitization Approach', *Human Security Journal* 5, 2007

'The Environmental Security Debate and its Significance for Climate Change', *The International Spectator* 43, 2008, 51–65

Friedman, Thomas, BBC2 *Newsnight*, 17 April 2008

Friends of the Earth Netherlands, Lembaga Gemawan Indonesia and KONTAK Rakyat Borneo, 'Policy, Practice, Pride and Prejudice', Amsterdam, Friends of the Earth Netherlands, 2007

Fritsch, Philippe, 'Einführung', in Franz Schultheiss and Luis Pinto (eds.), *Pierre Bourdieu: Das politische Feld: Zur Kritik der politischen Vernunft*, Konstanz, UVK Verlagsgesellschaft, 2001, 7–29

'Im Gespräch mit Pierre Bourdieu', in Franz Schultheiss and Luis Pinto (eds.), *Pierre Bourdieu: Das politische Feld: Zur Kritik der politischen Vernunft*, Konstanz, UVK Verlagsgesellschaft, 2001, 29–40

Funke, Odelia, 'Environmental Dimensions of National Security', in Jyrki Käkönen (ed.), *Green Security or Militarized Environment*, Aldershot, Dartmouth, 1994, 55–82

Gates, Robert, 'Speech to the Association of American Universities', Washington DC, Office of the Assistant Secretary of Defense, Public Affairs, 14 April 2008

George, Jim, *Discourses of Global Politics: A Critical (Re)Introduction to International Relations*, Boulder, Lynne Rienner, 1994

Giddens, Anthony, *The Politics of Climate Change*, Cambridge, Polity, 2009

Gleick, Peter H., *The World's Water: The Biennial Report on the World's Fresh Water Resources*, Washington DC, Island Press, 1989

Goldstone, Jack, 'Debate', *Environmental Change and Security Project Report*, Washington DC, The Woodrow Wilson Center, 1996, 66–71

Goodman, Sherri W., 'The Environment and National Security', speech at the National Defense University, Washington DC, Office of the Press Secretary, 8 August 1996

Gore, Albert, *Earth in Balance*, London, Earthscan Publications, 1992

Griffin, James, *Well-Being: Its Meaning, Measurement and Moral Importance*, Oxford, Clarendon Press, 1986

'The Human Good and the Ambitions of Consequentialism', *Social Philosophy and Policy* 9, 1992, 118–32

Gutting, Gary, *French Philosophy in the Twentieth Century*, Cambridge University Press, 2001

Halper, Stefan and Clarke, Jonathan, *America Alone: The Neo-Conservatives and the Global Order*, Cambridge University Press, 2004

Harris, Paul, 'Bush Covers up Climate Research', *The Observer*, 21 September 2003

Harris, Paul (ed.), *The Environment, International Relations and US Foreign Policy*, Washington DC, Georgetown University Press, 2001

Hartmann, Betsy, 'Rethinking the Role of Population in Human Security', in Richard A. Matthew, Jon Barnett, Bryan McDonald and Karin L. O'Brian (eds.), *Global Environmental Change and Human Security*, Cambridge MA, MIT Press, 2009, 192–214

Haspel, Abraham, Hecht, Alan and Vest, Gary, 'The DoD–DoE–EPA Environmental Security Plan', *Environmental Change and Security Project Report*, Washington DC, The Woodrow Wilson Center, 1997, 162–6

Holland, Steve, 'Obama Says Climate Change a Matter of National Security', *Reuters*, 9 December 2008

Homer-Dixon, Thomas F., 'On the Threshold: Environmental Changes as Causes of Acute Conflict', *International Security* 16, 1991, 76–116

'Environmental Scarcities and Violent Conflict: Evidence from Cases', *International Security* 19, 1994, 5–40

'Debate between Thomas Homer-Dixon and Marc A. Levy', *Environmental Change and Security Project Report*, Washington DC, The Woodrow Wilson Center, 1996, 49–60

Environment, Scarcity, and Violence, Princeton, Princeton University Press, 1999

The Ingenuity Gap, New York and Toronto, Alfred A. Knopf, 2000

Homer-Dixon, Thomas F. and Percival, Val, 'The Case of South Africa', in Paul F. Diehl and Nils Petter Gleditsch (eds.), *Environmental Conflict*, Oxford, Westview Press, 2001, 13–35

Huysmans, Jef, 'Language and the Mobilization of Security Expectations: The Normative Dilemma of Speaking and Writing Security', unpublished paper, presented at the ECPR Joint Sessions, Mannheim, Germany, 26–31 March 1999

Inhofe, James, 'The Science of Climate Change Senate Floor Statement', US Senate Committee on Environment and Public Works, Washington DC, 28 July 2003

Jackson, Richard, *Writing the War on Terrorism: Language, Politics and Counter-Terrorism*, Manchester, Manchester University Press, 2005

Jones, Richard Wyn, *Security, Strategy, and Critical Theory*, Boulder, Lynne Rienner, 1999

'On Emancipation: Necessity, Capacity, and Concrete Utopias', in Ken Booth (ed.), *Critical Security Studies and World Politics*, Boulder, Lynne Rienner, 2005, 215–36

Käkönen, Jyrki (ed.), *Green Security or Militarized Environment*, Aldershot, Dartmouth, 1994

Kaplan, Robert, 'The Coming Anarchy', in Gearóid Ó Tuathail, Simon Dalby and Paul Routledge, *The Geopolitics Reader*, London, Routledge, 1998, 188–96

Keane, General John, 'The Impact of Environmental Extremism on Military Readiness: The Encroachment Problem', Washington DC, US Senate Republican Committee, 2003

Kegley, Charles W., Wittkopf, Eugene R. and Scott, James M., *American Foreign Policy*, sixth edition, London, Thomson Wadsworth, 2003

Kingwell, Mark, 'Meet Tad The Doom-meister' *Saturday Night*, September 1995, 43–6

Klare, Michael T., *Resource Wars: The New Landscape of Global Conflict*, New York, Henry Holt and Company, 2001

Kurki, Milja, 'Causes of a Divided Discipline: Rethinking the Concept of Cause in International Relations', *Review of International Studies* 32, 2006, 189–216

   Causation in International Relations: Reclaiming Causal Analysis, Cambridge University Press, 2008

Langer, Gary, 'Still Proud to be an American – Poll: One Year Later Public Remains Proud, Optimistic despite Fears', *ABC News*, 10 September 2002

Lanier-Graham, Susan D., *The Ecology of War: Environmental Impacts of Weaponry and Warfare*, New York, Walker and Company, 1993

Lemann, Nicholas, 'Dreaming about War', *New Yorker*, 16 July 2001

Levy, Marc A., 'Is the Environment a National Security Issue?' *International Security* 20, 1995, 35–62

Lippman, Thomas, 'With Tim Wirth in Position, The Old Lines Lose Weight', *Washington Post*, 30 June 1994

   'Tim Wirth versus State', *Washington Post*, 20 April 1995

   'Christopher puts Environment at Top of Diplomatic Agenda', *Washington Post*, 15 April 1996

Lonergan, Steve C., 'Water and Conflict: Rhetoric and Reality', in Paul F. Diehl and Nils Petter Gleditsch (eds.), *Environmental Conflict*, Oxford, Westview Press, 2001, 109–24

Lowi, Miriam R. and Shaw, Brian R. (eds.), *Environment and Security: Discourses and Practices*, New York, St Martin's Press, 2002

MacAskill, Ewan, 'Europeans Angry after Bush Climate Speech Charade', *The Guardian*, 29 September 2007

MacFarlane, S. Neil and Yuen Foong Kong, *Human Security and the UN: A Critical History*, Indianapolis, Indiana University Press, 2006

Martinson, Jane, 'Poll Shows Half of Americans Doubt Bush's Trustworthiness: Special Report on George Bush's America', *The Guardian*, 28 May 2001

Mathews, Jessica Tuchman, 'Redefining Security', *Foreign Affairs* 68, 1989, 162–77

McGray, Douglas, 'The Marshall Plan', *Wired*, 2003 www.wired.com/wired/archive/11.02/marshall.html

McNaughton, David and Rawling, Piers, 'Honoring and Promoting Values', *Ethics* 102, 1992, 835–43

Meyer, Dan and Volk, Everett E., 'W for War or Wedge? Environmental Enforcement and the Sacrifice of American Security – National and

Environmental – to Complete the Emergence of a New "Beltway" Elite', *Western New England Law Review* 25, 2003, 41–146

Monbiot, George, 'The Planet is Now so Vandalized that only Total Energy Renewal Can Save Us', *The Guardian*, 25 November 2008

Mulgan, Tim, *The Demands of Consequentialism*, Oxford, Clarendon Press, 2001

Myers, Norman, 'The Environmental Dimension to Security Issues', *The Environmentalist* 6, 1986, 251–7

'Environment and Security', *Foreign Policy* 74, 1989, 23–41

Ultimate Security: The Environmental Basis of Political Stability, New York, Norton, 1993

Myers, Norman and Simon, Julian, *Scarcity or Abundance? A Debate about the Environment*, London, W.W. Norton & Company, 1994

Nagel, Thomas, 'War and Massacre', in Samuel Scheffler (ed.), *Consequentialism and its Critics*, Oxford, Oxford University Press, 1988, 51–73

Nitze, William, 'A Potential Role for the Environmental Protection Agency and Other Agencies', *Environmental Change and Security Project Report*, Washington DC, The Woodrow Wilson Center, 1996, 116–20

Northwest Florida Greenway Memorandum of Partnership among Department of Defense, State of Florida, and the Florida Chapter of the Nature Conservancy to Conserve Environmentally Significant Lands and Limit Incompatible Development in Northwest Florida, 2003

Office of Management and Budget, 'Budget of United States Government FY2008', Washington DC, Department of Defense, 2008

Office of the Deputy Secretary of Defense, *Sustainable Ranges 2003 Decision Briefing to the Deputy Secretary of Defense*, Washington DC, Department of Defense, 2002

'Memorandum for the Secretary of the Army, the Navy and the Air Force; Subject: Consideration of Requests for use of Existing Exemptions under Federal Environmental Laws', Washington DC, Department of Defense, 2003

Office of the Deputy Under Secretary of Defense (Environmental Security), '1994 Defense Environmental Quality (EQ) Program Annual Report to Congress', Washington DC, Department of Defense, 1994

'1994 Defense Environmental Restoration Programs (DERP) Annual Report to Congress', Washington DC, Department of Defense, 1994

'Mission Statement Environmental Security Education', Washington DC, Department of Defense, 1994

Report to the Defense Science Board Task Force on Environmental Security, Washington DC, Department of Defense, 1995

'1996 Defense Environmental Quality (EQ) Programs Annual Report to Congress', Washington DC, Department of Defense, 1996

'2000 Defense Environmental Quality (EQ) Programs Annual Report to Congress', Washington DC, Department of Defense, 2000

Office of the Deputy Under Secretary of Defense (Installations and Environment), '2001 Defense Environmental Quality (EQ) Programs Annual Report to Congress', Washington DC, Department of Defense, 2001

'2002 Defense Environmental Quality (EQ) Programs Annual Report to Congress', Washington DC, Department of Defense, 2002

'Pollution Prevention in Progress Review', Washington DC, Department of Defense, 2002

'2004 Defense Environmental Programs (DEP) Annual Report to Congress', Washington DC, Department of Defense, 2004

Department of Defense Sustainable Ranges: Better Planning through Partnership, Washington DC, Department of Defense, 2005

'2007 Defense Environmental Programs (DEP) Report to Congress', Washington DC, Department of Defense, 2007

'2008 Defense Environmental Programs (DEP) Annual Report to Congress', Washington DC, Department of Defense, 2008

Office of the Secretary of Defense, 'Assured Fuels Initiative Slide Show', Washington DC, Department of Defense, 2006

Office of the Under Secretary of Defense for Acquisition, Technology, and Logistics, Department of Defense Directive 4715.1E, 'Environment, Safety, and Occupational Health (ESOH)', Washington DC, Department of Defense, 19 March 2005

Oxfam Briefing Note, *Bio-fuelling Poverty: Why the EU Renewable-fuel Target may be Disastrous for Poor People*, Oxford, Oxfam International, 2007

Palmer, Clare, 'An Overview of Environmental Ethics', in Andrew Light and Holmes Rolston III (eds.), *Environmental Ethics: An Anthology*, Oxford, Blackwell, 2003, 15–37

Parfit, Derek, *Reasons and Persons*, Oxford, Clarendon Press, 1984

Paterson, Matthew, *Understanding Global Environmental Politics: Domination, Accumulation and Resistance*, Basingstoke, Palgrave, 2001

Peluso, Nancy L. and Watts, Michael, *Violent Environments*, New York, Cornell, 2001

Pemberton, Miriam, *The Budgets Compared: Military vs. Climate Security*, Washington DC, Institute for Policy Studies, 2008

Perry, William J., 'Good Stewards at Home, Good Stewards Abroad', Remarks to John F. Kennedy School of Government, Harvard University, Washington DC, Office of the Press Secretary, 13 May 1996

Pettit, Philip, 'Introduction', in Philip Pettit (ed.), *Consequentialism*, Aldershot, Dartmouth Press, 1993, xiii–xix

Phase II report on a US National Security Strategy for the 21st century 'Seeking a National Strategy: A Concert for Preserving Security and for Promoting Freedom', Washington DC, US Commission on National Security 21st Century, 2000

Pirages, Dennis, 'Demographic Change and Ecological Security', *Environmental Change and Security Project Report*, Washington DC, The Woodrow Wilson Center 1997, 37–46

Pope, Carl and Rauber, Paul, *Strategic Ignorance: Why the Bush Administration is Recklessly Destroying a Century of Environmental Progress*, San Francisco, Sierra Club Books, 2004

Portmore, Douglas W., 'Can an Act-consequentialist Theory be Agent Relative?' *American Philosophical Quarterly* 38, 2001, 363–77

Powell, Colin, 'Making Sustainable Development Work: Governance, Finance and Public–Private Co-operation', Washington DC, Office of the Secretary of State, 2002

Price, Richard M., 'Moral Limit and Possibility in World Politics', in Richard M. Price (ed.), *Moral Limit and Possibility in World Politics*, Cambridge University Press, 2008, 1–52

Prins, Gwyn, *Threats Without Enemies: Facing Environmental Insecurity*, London, Earthscan Publications, 1993

Prins, Gwyn and Stamp, Robbie, *Top Guns and Toxic Whales: The Environment and Global Security*, London, Earthscan Publications, 1991

Public Employees for Environmental Responsibility, 'Sustainable Defense Readiness and Environmental Protection Act Discussion Draft', discussion draft not for release, Washington DC, 7 March 2003

'US Army Restores Environmental Funding', 28 May 2004

Railton, Peter, 'Alienation, Consequentialism and Morality', reprinted in Samuel Scheffler (ed.), *Consequentialism and its Critics*, Oxford, Oxford University Press, 1988, 93–133

Rawls, John, *A Theory of Justice*, Cambridge MA, Harvard University Press, 1971

Reus-Smit, Christian, 'Constructivism and the Structure of Moral Reasoning', in Richard M. Price (ed.), *Moral Limit and Possibility in World Politics*, Cambridge University Press, 2008, 53–82

Rice, Condoleezza, 'Dr Condoleezza Rice discusses President's National Security Strategy', Washington DC, Office of the Secretary of State, 1 October 2002

Rosenbaum, Walter A., *Environmental Politics and Policy*, fifth edition, Washington DC, Congressional Quarterly Press, 2002

Ruch, Jeff, 'Pentagon Files For Environmental Exemptions: Mixed Earth Day Message for Bush Administration', Washington DC, Public Employees for Environmental Responsibility, 2003

Rudolph, Dieter, 'Arctic Military Environmental Cooperation AMEC: An Effort of the International Defense Community to Protect the Coastal and Marine Environment in the Russian Arctic', unpublished manuscript, 2004

Rumsfeld, Donald H., 'Transforming the Military', *Foreign Affairs*, 81, 2002, 20–32

'Remarks at the White House Conference on Cooperative Conservation', Washington DC, Office of the Assistant Secretary of Defense, Public Affairs, 29 August 2005

Sachs, Jeffrey D., 'The Strategic Significance of Global Inequality', *Environmental Change and Security Project Report*, Washington DC, The Woodrow Wilson Center, 2003, 27–34

Sarkesian, Sam C., Williams, John Allan and Cimbala, Stephen, *US National Security Policymakers, Processes and Politics*, third edition, Boulder, Lynne Rienner, 2002

Schmitt, Carl, *The Concept of the Political*, Chicago, University of Chicago Press, 1996

Shulman, Seth, *The Threat at Home: Confronting the Toxic Legacy of the US Military*, Boston, Beacon Press, 1992

Sigler, John F., 'US Military and Environmental Security in the Gulf Region', *Environmental Change and Security Project Report*, Washington DC, The Woodrow Wilson Center, 2005, 51–7

Singer, Marcus G., 'Actual Consequence Utilitarianism', *Mind* 86, 1977, 67–77

Singer, Peter, *Practical Ethics*, second edition, Cambridge University Press, 1993

Skinner, Quentin, *Visions of Politics I. Regarding Method*, Cambridge University Press, 2002

Slote, Michael A., *Common-sense Morality and Consequentialism*, London, Routledge & Kegan Paul, 1985

Smith, Richard, 'The Intelligence Community and the Environment: Capabilities and Future Missions', *Environmental Change and Security Project Report*, Washington DC, The Woodrow Wilson Center, 1996, 103–8

Smoke, Richard, *National Security and the Nuclear Dilemma: An Introduction to the American Experience in the Cold War*, third edition, New York, McGraw-Hill, 1993

Smythe, Robert, Author interview, Washington DC, 17 September 2005

Stritzel, Holger, 'Towards a Theory of Securitization: Copenhagen and Beyond', *European Journal of International Relations* 13, 2007, 357–83

Symons, Jeremy, 'How Bush and Co. Obscure the Science', *Washington Post*, 13 July 2003

Tatalovich, Raymond and Wattier, Mark, J., 'Opinion Leadership: Elections, Campaign, Agenda Setting, and Environmentalism', in Dennis L. Soden (ed.), *The Environmental Presidency*, New York, State University of New York Press, 1999, 147–87

Taureck, Rita and Dabelko, Geoffrey D., 'Profile of the United States', in Ronald A. Kingham (ed.), *Inventory of Environment and Security Policies and Practices: An Overview of Strategies and Initiatives of Selected Governments, International Organisations and Inter-Governmental Organisations*, The Hague, Institute for Environmental Security, 2006, 113–22

Thomas, Kenneth, 'Official Statement, Department of State/Office of Under Secretary for Global Affairs', *Environmental Change and Security Project Report*, Washington DC, The Woodrow Wilson Center, 1995, 84

Ullman, Richard, 'Redefining Security', *International Security* 8, 1983, 129–53

UN Energy, 'Sustainable Bio-energy: A Framework for Decision Makers', New York, United Nations, 2007

UN Human Development Report 2007/2008, *Fighting Climate Change: Human Solidarity in a Divided World*, New York, United Nations Development Programme, 2007

USAID Budget Justifications FY2001–FY2009 www.usaid.gov/policy/budget/

US Department of Defense, Directive DOD 4715.1 Environmental Security, 24 February 1996

Instruction Number 4715.6 Environmental Compliance, 24 April 1996

Instruction Number 4715.9 Environmental Planning and Analysis, 3 May 1996

Directive 4700.4 Natural Resources Management Program, 24 January 1998

'Quadrennial Defense Review Report', Washington DC, Department of Defense, 2001

DOD Sustainable Ranges Initiative, Washington DC, internal publication courtesy of the Office of the Under Secretary of Defense (Installations and Environment), 2005

'Environment, Safety, and Occupational Health (ESOH) Directive', 19 March 2005

'Quadrennial Defense Review Report', Washington DC, Department of Defense, 2006

'2008 National Defense Strategy', Washington DC, Department of Defense, June 2008

US Department of Energy, 'The Energy Policy Act of 2005 Tax Credits', Washington DC, Department of Energy, 2005

US Department of State, *Environmental Diplomacy: The Environment and US Foreign Policy*, Washington DC, Department of State, 1997

US Environmental Protection Agency, *Environmental Security: Strengthening National Security through Environmental Protection*, Washington DC, Environmental Protection Agency, 1999

US Fish and Wildlife Service, 'Integrated Natural Resources Management Plan Fact Sheet', Washington DC, Department of the Interior, 2004

United States General Accounting Office, *Military Lacks a Comprehensive Plan to Manage Encroachment on Training Ranges* (02-614), Washington DC, US General Accounting Office, 2002

United States House of Representatives, 'National Defense Authorization Act for Fiscal Year 2008', Washington DC, US Congress, 2008

United States National Security Strategy 1991, 'The National Security Strategy of the United States', Washington DC, The White House, August 1991

United States National Security Strategy 1995, 'A National Security Strategy of Engagement and Enlargement, Washington DC, The White House, 1995

United States National Security Strategy 1999, 'A National Security Strategy for a New Century', Washington DC, The White House, December 1999

United States National Security Strategy 2002, 'National Security Strategy of the United States of America', Washington DC, The White House, September 2002

United States National Security Strategy 2006, 'The National Security Strategy of the United States of America', Washington DC, The White House, March 2006

Vest, Gary D. and Lietzmann Kurt M., *Environment and Security in an International Context*, Washington DC, Department of Defense, 1999

Wæver, Ole, 'Beyond the "Beyond" of Critical International Theory', unpublished paper, presented at the Joint Annual Convention of the British International Studies Association and the International Studies Association, London, 1989

'Security, the Speech Act: Analysing the Politics of a Word', unpublished paper, presented at the Research Training Seminar, Sostrup Manor, 1989, revised, Jerusalem/Tel Aviv 25–26 June 1989

'Tradition and Transgression in International Relations as Post-Ashleyan Position' unpublished paper, presented at the British International Studies Association 15th Annual Conference at the University of Kent, 1989

'Securitization and Desecuritization', in Ronnie D. Lipschutz (ed.), *On Security*, New York, Columbia University Press, 1995, 46–86

Concepts of Security, Copenhagen, Institute of Political Science, University of Copenhagen, 1997

'Securitizing Sectors? Reply to Eriksson', *Cooperation and Conflict* 34(3), 1999, 334–40

'Security Agendas Old and New and How to Survive Them', unpublished paper, presented at the workshop on *'The Traditional and the New Security Agenda: Inferences for the Third World'*, Universidad Torcuato di Tella, Buenos Aires, 11–12 September 2000

'Identity, Communities and Foreign Policy', in Lene Hansen and Ole Wæver (eds.), *European Integration and National Identity: The Challenge of the Nordic States*, London, Routledge, 2001, 20–49

'Security: A Conceptual History for International Relations', unpublished paper, presented at the British International Studies Association 29th Annual Conference in London, 2002

'Securitisation: Taking Stock of a Research Programme in Security Studies', unpublished manuscript, 2003

'Aberystwyth, Paris, Copenhagen: New Schools in Security Theory and the Origins between Core and Periphery', unpublished paper, presented at the International Studies Association 45th Annual Convention in Montreal, Canada, 2004

'The Ten Works', *Tidsskriftet Politik* 7, 2004

Wæver, Ole, Buzan, Barry, Kelstrup, Morten and Lemaitre, Pierre, *Identity, Migration and the New Security Agenda in Europe*, London, Pinter, 1993

Walker, R. B. J., *Inside/Outside: International Relations as Political Theory*, Cambridge University Press, 1993

Walker, Gabrielle and King, Sir David, *The Hot Topic: How to Tackle Climate Change and Still Keep the Lights On*, London, Bloomsbury, 2008

Waltz, Kenneth, *Theory of International Politics*, New York, Random House, 1979

Wendt, Alexander, *Social Theory of International Relations*, Cambridge University Press, 1999

Westing, Arthur H., *Warfare in a Fragile Environment: Military Impact on the Human Environment*, London, Taylor and Francis, 1980

'Environmental Warfare: Manipulating the Environment for Hostile Purposes', *Environmental Change and Security Project 3*, Washington DC, The Woodrow Wilson Center, 1997, 145–9

(ed.) *Herbicides in War: The Long Term Ecological and Human Consequences*, London, Taylor and Francis, 1984

Williams, Michael C., 'Words, Images, Enemies: Securitization and International Politics' *International Studies Quarterly* 47, 2003, 511–31

Culture and Security: Symbolic Power and the Politics of International Security, Abingdon, Routledge, 2007

Wirth, Timothy E., *Sustainable Development and National Security*, Washington DC, Bureau of Public Affairs, US Department of State, 1994
  Environmental Challenges Confront the Post-Cold War World, Washington DC, Office of the Under Secretary of State for Global Affairs, 1995
Wolf, Aaron, '"Water Wars" and Water Reality: Conflict and Cooperation along International Waterways', in Steve C. Lonergan (ed.), *Environmental Change, Adaptation, and Security*, Dordrecht, Kluwer Academic Publishers, 1999, 251–65

Index

35766226R00130

Printed in Poland
by Amazon Fulfillment
Poland Sp. z o.o., Wrocław